Volume
1

PRANIT KUMAR SAHA

Oracle Magic
With
Forms Personalizations

Dedicated to my idol – Parents (Mr. Paresh Nath Saha and Mrs. Gita Saha) and my motivation – Sons (Shriyan and Pranil) and Wife (Soumi).

Oracle Magic with Forms Personalizations

21B Hope Lane
Mawson Lakes, South Australia 5095
Phone +61 426254834.

Table of Contents

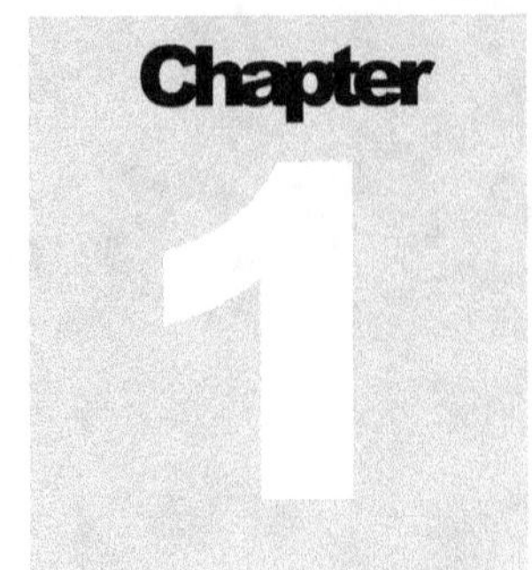

Forms Personalization Overview

This chapter provides an overview of Forms Personalization and discusses about various components of the Personalization window. You will be able to understand the utilities and importance of the key Forms Personalization objects.

There are many scenarios where we, the consultants often come to conclusion that customer requirements, such as – ability to search stock based on manufacturer part number, exporting data from all the tabs of “Material Transactions” form to a spreadsheet are not possible in Oracle Applications or require a major customization. On the other hand, users of the Oracle Applications often complaints about the number of key strokes or mouse clicks they have to perform while entering data into the form.

Oracle Forms Personalization is a magical capability and a declarative feature that has been provided in e-Business Suite version 11.5.10 onwards that allows us to extend the application capability significantly to deliver those niche customer requirements and improve user experience by modifying the behavior of the Oracle Forms.

The main objective of writing this book is to help the consultants understand the power of Forms Personalization which will help them deliver much effective and efficient solution for the business by avoiding expensive customizations and improving user experience.

Points to be noted:

- Only the Trigger Events processed by APPCORE can be used for Forms Personalization.
- Not all the form objects will be available for Personalization.
- Seeded form code might override the Personalization effect; hence the changes may not be always visible.
- Ability to perform Personalization should be available only to the authorized users.
- The “Utilities: Diagnostics” profile option needs to be setup as “Yes” at the respective user level to access the form personalization window through Help -> Diagnostics -> Custom Code -> Personalize menu option.
- It is always recommended to perform in-depth testing of the Personalization in the development and Testing environment prior to migrating the Personalization to the Production environment.
- Examples mentioned in this book are based on Author’s own experience and opinion. The Author is not responsible for any damage or negative impact caused while performing example scenarios and steps on your system.

Underlying Tables:

Following tables store the Personalization configuration details.

- FND_FORM_CUSTOM_RULES
- FND_FORM_CUSTOM_ACTIONS

- FND_FORM_CUSTOM_SCOPES
- FND_FORM_CUSTOM_PROP_LIST
- FND_FORM_CUSTOM_PROP_VALUES
- FND_FORM_CUSTOM_PARAMS
- FND_INDUSTRIES

Components of Forms Personalization:

- Debug mode: As the name suggests, this helps to debug the forms personalization issues to identify the root cause. There are 2 Debug Modes as mentioned below.
    - Show Debug Messages: This option is useful to debug a personalization step by displaying a user defined debug message using the "Debug" message type (Discussed later in this book). This is particularly helpful when you are dealing with local and global variables in the Forms Personalization to understand whether the variables are properly assigned.
    - Step-by-Step: As Oracle processes the personalization rules sequentially based on an ascending order, the developers often find that the some of the defined rules are not getting activated. This is usually because the higher level rules having superseded the lower level ones. In order to understand the processing pattern and identify the root cause of a problem accordingly, the "Step-by-Step" debug mode can be enabled. This will prompt a message in each step to help the developer understand the trigger/event it's currently processing. Below screen shot is an example of such messages.

    The key difference between "Show Debug Messages" and "Step-by-Step" Debug Mode options is – the former shows the user defined debug messages while the latter shows the system defined messages.

- Level: As we know, the forms are attached to the functions and functions are attached to the menus which are accessible through various responsibilities. There are many forms in the Oracle Applications which are attached to multiple functions. For example, the "Master Items" and

"Organization Items" menus available in the Inventory Superuser or similar responsibilities are actually opening the form INVIDITM but the menus are referring to 2 different functions, as the form behaves differently in the mentioned menu items. The "Master Items" menu is used to create items while the "Organization Items" menu allows users to query and update item attributes. If there is a personalization that auto generates an Item number as soon as the form is launched then this needs to be active for the "Master Items" menu but not for the "Organization Items" menu. Therefore the mentioned personalization should be configured with the level value as "Function". On the other hand, when a personalization is applicable to the form regardless of the function, then the level should be defined as "Form"

- Condition – This section of the form personalization allows developer to define the business logic and it includes the following sections.

    - Trigger Event – Provided a list of trigger events where the business logic to be validated.

    - Trigger Object – Provides a list of objects based on the "Trigger Event" selected in the previous step.

    - Condition – The business logic needs to be mentioned in this field which will be validated by the "Trigger Event" selected earlier. This is an optional field and the value might look like – **${item.mtl_system_items.inventory_item_mir.value} IS NULL**.

    The syntax to be entered in the "Condition" field can be obtained using the "Insert 'Get' Expression" and "Insert Item Value" button.

    You will get the below form to generate the syntax when you click on the "Insert 'Get' Expression" button.

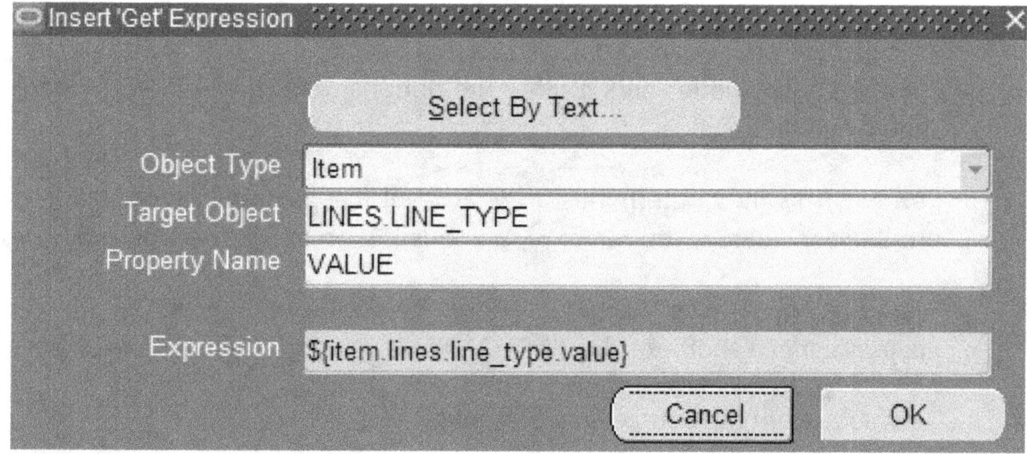

The object type, Target Object and Property Name needs to be fulfilled and that will generate the syntax in the Expression field.

You may also use the “Select By Text” button to select the object based on the field name, as this is much user-friendly option to identify a Target Object.

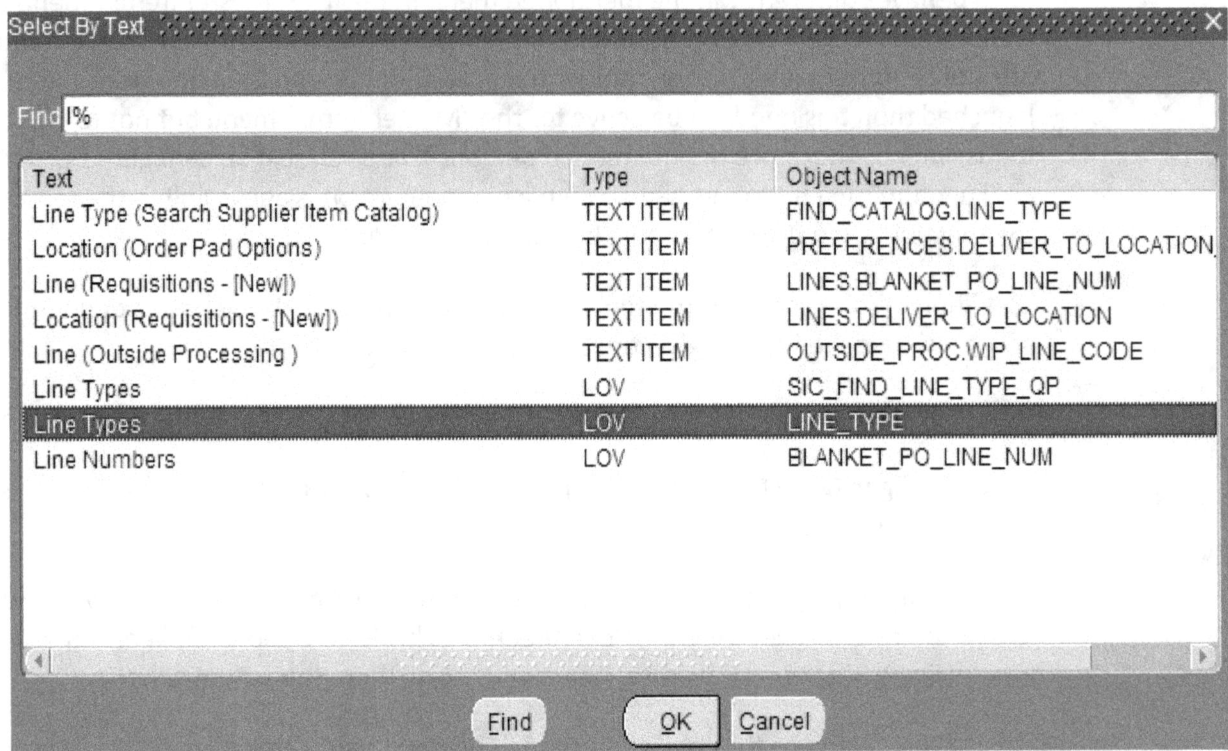

On the other hand, “Insert Item Value” button generates a syntax based on “:block.field” structure. Therefore the syntax generated in the previous option will look like “:LINES.LINE_TYPE”.

- Processing mode: There are 3 options in this field as – “Not in Enter-Query Mode”, “Only in Enter-Query Mode” and “Both”. The options determine when the personalization rule should be processed.

- Context: This field determines at what level the personalization is applicable. Similar to profile option setup, you can apply personalization at Site, Responsibility or User level.

- Migration of Forms Personalization: Personalization can be migrated from one environment to another using FNDLOAD utility and the syntax is as follows.

  Download from the Source environment:

  FNDLOAD <user>/<password> O Y DOWNLOAD $FND_TOP/patch/115/import/affrmcus.lct <filename.ldt> FND_FORM_CUSTOM_RULES FUNCTION_NAME=<function_name>

Upload to the Destination environment:

FNDLOAD <user>/<password> O Y UPLOAD $FND_TOP/patch/115/import/affrmcus.lct <filename.ldt>

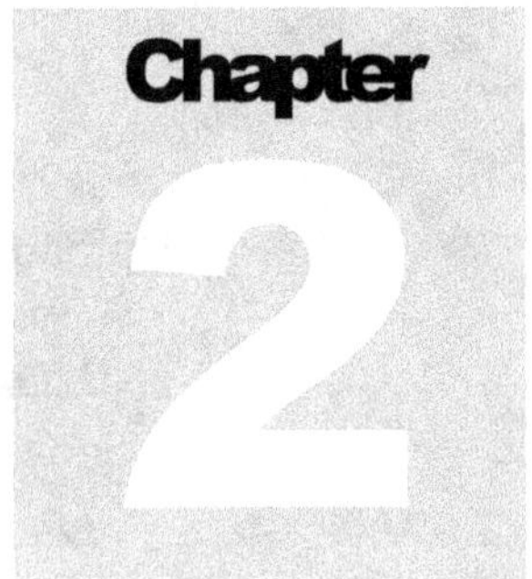

Triggers

This chapter explains various triggers supported by Forms Personalization. You will be able to understand the utilities and execution points of each supported trigger.

WHEN-NEW-FORM-INSTANCE:

This trigger event is fired when a form is opened in response to clicking/double-clicking a menu item in the Oracle Applications. This may be used to fulfil following business requirements:

- Prompt a message as soon as form is opened.
- Default a value in a field.
- Data sorting
- Enable/disable a form field.
- Show/Hide a form field.
- Change physical attributes (height, width etc.) of a canvas/window etc.

We will discuss about each actions that can be performed using forms personalizations in the subsequent chapters in this book.

WHEN-NEW-BLOCK-INSTANCE

As we may already know, each form contains one or many blocks and the fields are nested inside the block. WHEN-NEW-BLOCK-INSTANCE trigger is fired when the focus of an activity is moved to a block. For example: most of the forms (e.g. Oracle Requisition form) have a Header section and detail section. Usually these sections are blocks. In addition to the activities mentioned in the previous trigger event, you will be able to perform additional validations on a block based on the data from the other block (e.g. validation on detail block based on the data from header block).

How to identify the block name to be entered as Trigger Object?

Best way to identify the block is to query a record in the form, place your cursor in one of the value in the respective block and then click on Help -> Diagnostics -> Examine menu item to identify the Block name. Following example will clarify the same further.

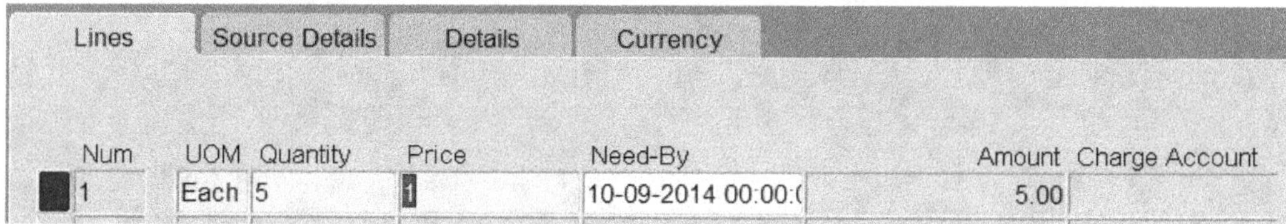

The above screen shot is from the Requisition line level. Now if you want to perform some personalisations here, you may need to know the block name in order to mention in the personalization setup form. Please perform the following steps to identify the block:

- Place your cursor in one of the field where data is shown (I have placed my cursor in the price field as shown above).
- Click on the Help -> Diagnostics -> Examine menu

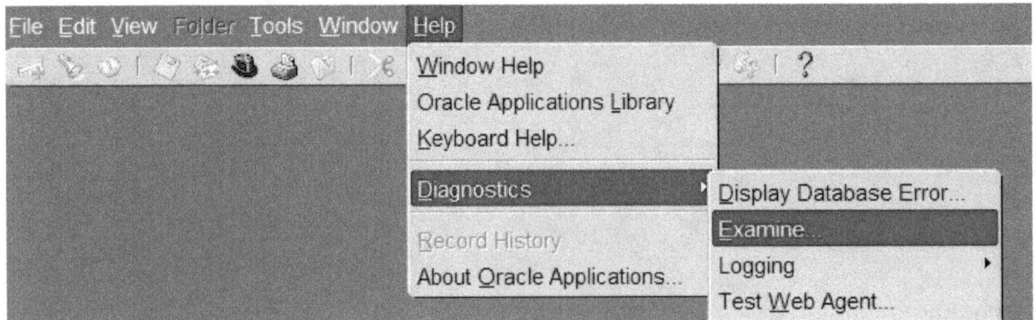

- You will see the following information that includes block name.

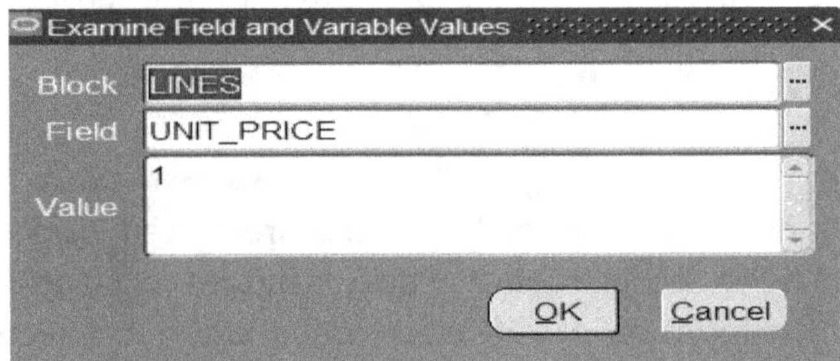

The WHEN-NEW-BLOCK-INSTANCE trigger is fired after the WHEN-NEW-FORM-INSTANCE trigger.

WHEN-NEW-RECORD-INSTANCE

This trigger is fired when you create a new line/record in the block. For example: If you are creating line 2 for the requisition shown in the below screen shot and you want to default the need by date as SYSDATE+5 as soon as the focus moves to line 2, you may create a personalization in the WHEN-NEW-RECORD-INSTANCE trigger to fulfil the requirement.

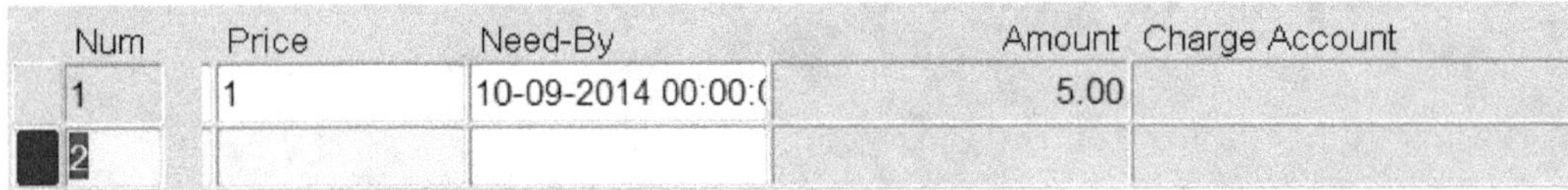

In the above screen shot, the WHEN-NEW-RECORD-INSTANCE trigger is fired as soon as the cursor moves to line 2 (as highlighted).

You need to select the block name as Trigger Object to setup the personalization for WHEN-NEW-RECORD-INSTANCE trigger as shown below.

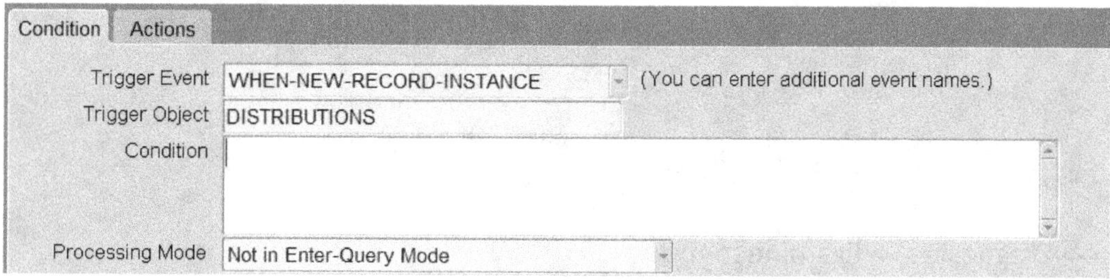

WHEN-NEW-RECORD-INSTANCE trigger is fired following the below sequence:

WHEN-NEW-FORM-INSTANCE -> WHEN-NEW-BLOCK-INSTANCE -> WHEN-NEW-RECORD-INSTANCE

<u>WHEN-NEW-ITEM-INSTANCE</u>

WHEN-NEW ITEM-INSTANCE trigger is fired at the field level. That means, based on the below screen shot, this trigger may be fired at fields such as Number, Price, Need-By date etc.

Num	Price	Need-By	Amount	Charge Account
1	1	10-09-2014 00:00:(	5.00	
2				

For example, if the business requirement is to default SYSDATE+10 in the Need-By date column as well as make it read-only/un-editable so that user can't change it, then you may create a form personalization by firing WHEN-NEW-ITEM-INSTANCE trigger at the Need-By field. More details about this type of personalization will be discussed in subsequent chapters of this book.

You need to select the appropriate item (as shown below) as trigger object to setup this personalization to indicate the field from where this trigger will be fired.

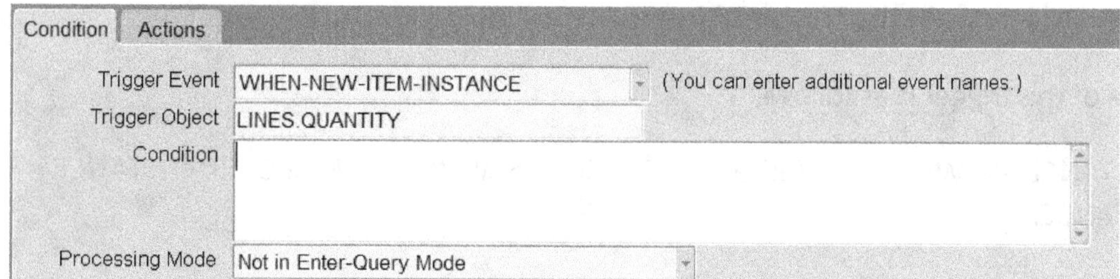

<u>How to identify the item entered as Trigger Object?</u>

Similar to the steps explained above to identify the block in the WHEN-NEW-BLOCK-INSTANCE trigger section, you need to follow below steps.

- Place your cursor in the field where you are planning to fire the trigger from (I have placed my cursor in the price field as shown below).

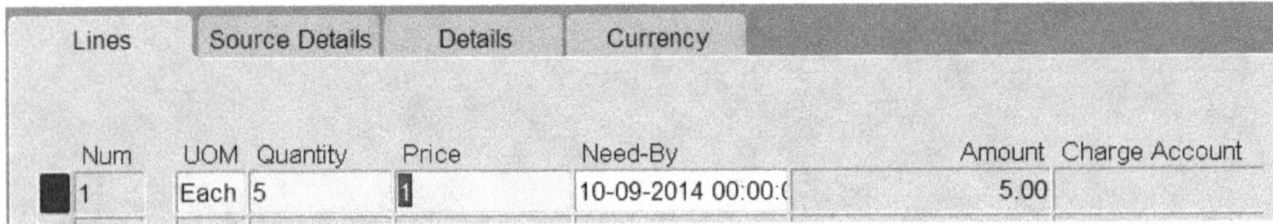

- Click on the Help -> Diagnostics -> Examine menu

- You will see the following information that includes block and field name.

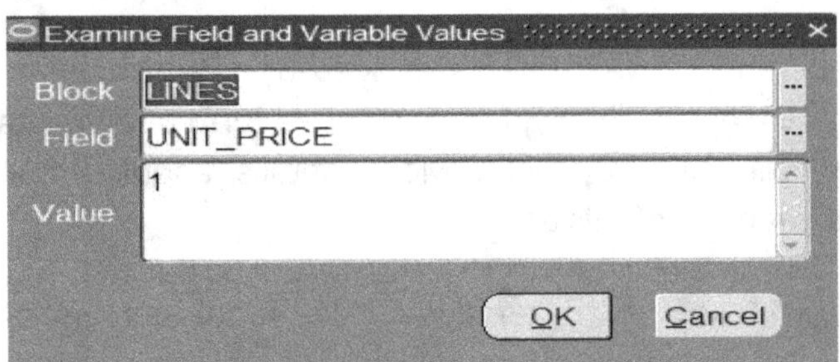

The field value "UNIT_PRICE" is actually the Trigger Object in this example.

The execution sequence of this trigger is as follows:

WHEN-NEW-FORM-INSTANCE -> WHEN-NEW-BLOCK-INSTANCE -> WHEN-NEW-RECORD-INSTANCE -> WHEN-NEW-ITEM-INSTANCE.

WHEN-VALIDATE-RECORD

As the name suggests, this trigger is usually fired to validate the entered record unlike the above mentioned triggers which is usually used for defaulting values and control attribute characteristics (e.g. Enabled (Yes/No), Visible (Yes/No) etc.

WHEN-VALIDATE-RECORD trigger is usually fired prior to saving the record to the tables; therefore this is very effective if you want to validate the data entered into a block and throw error messages as appropriate.

For example, referring to the PO requisition form, if the line type is "Goods" and UOM is "HR" (Hour) the business may require to restrict such requisition lines and throw an error message accordingly, as Hours based lines are usually created with "Services" line type. To elaborate this further, if you are creating a requisition to hire an Oracle Functional consultant for 1000 Hours, the line type must be selected as "Services", as this kind of professional service isn't tangible; hence not a Goods.

WHEN-VALIDATE-RECORD trigger can be effectively used in the above mentioned business scenario to check the line type and UOM field values and throw an error message such as "Please select appropriate line type."

You need to select the appropriate block to setup forms personalization for this trigger as shown below.

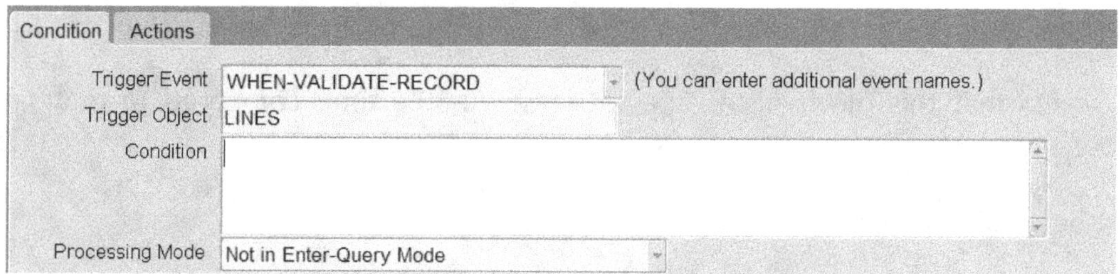

The execution sequence of this trigger is as follows:

WHEN-NEW-FORM-INSTANCE -> WHEN-NEW-BLOCK-INSTANCE -> WHEN-NEW-RECORD-INSTANCE -> WHEN-NEW-ITEM-INSTANCE. -> WHEN-VALIDATE-RECORD.

MENUn and SPECIAL n

We have seen there are handful of menus available with each form in Oracle e-Business Suite but frequently we come across the need of having additional menus so that we can make Oracle more user-friendly. Let's discuss on the below example and see how can we use MENU or SPECIAL personalizations.

Example: You are creating a Purchase Order using the Purchase Order form and now you would like to see how the PDF copy looks like prior to submitting it for approval.

In this case, you need to submit the "PO Output for Communication" program manually by clicking on View-> Request, "Submit a New Request" and then "Single Request" etc. and subsequently you also need to enter the PO number and other parameter values manually prior to submitting the program. All that takes time and imagine if you are a buyer who creates at least 50-60 purchase orders a day and would like to review the PDF for all of them prior to submitting it for approval; you better have a user-friendly mechanism to be more productive.

What we can do is, provide a sub-menu under the "Tools" menu as "Launch PDF", so that the buyer can launch the program directly from the Purchase Order form in a couple of clicks as well as program can be auto-populated with the critical parameter values. The Buyer just needs to review the parameter values and submit the program.

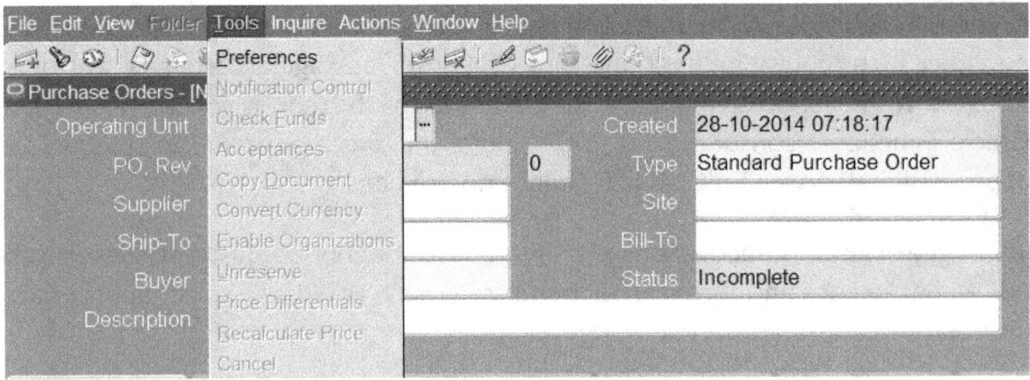

The above screen shot shows the seeded submenus under Tools menu and we are suggesting addition of "Launch PDF" submenu there.

A rough time-study indicates that this mechanism saves at least 70-80% time compared to manual program submission process.

Following steps explains the details of the proposed personalization.

Step-1:

Define the sub-menu.

Function Name PO_POXPOEPO Form Name POXPOEPO Debug Mode Off

Seq	Description	Level	Enabled
10	Define "Launch PDF" submenu	Function	☑

Condition | Actions

Trigger Event WHEN-NEW-FORM-INSTANCE (You can enter additional event names.)

Trigger Object

Condition

Processing Mode Not in Enter-Query Mode

Seq – 10

Description – Define "Launch PDF" submenu.

Level – Function

Enabled – Yes

Condition:

Trigger Event: WHEN-NEW-FORM-INSTANCE (Note: Sub menus can be defined only in the WHEN-NEW-FORM-INSTANCE trigger.)

Condition: <You may enter your condition here>

Processing Mode: Not in Enter-Query Mode.

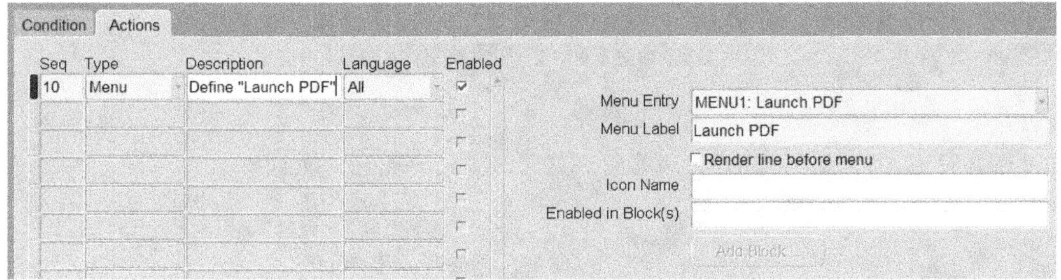

Actions:

Seq: 10

Type: Menu

Description: Define "Launch PDF"

Language: All

Enabled: Yes

Menu Entry: MENU1

Menu Label: Launch PDF

Now you have completed the setups required to create the submenu "Launch PDF" under the Tools menu as per below screen shot.

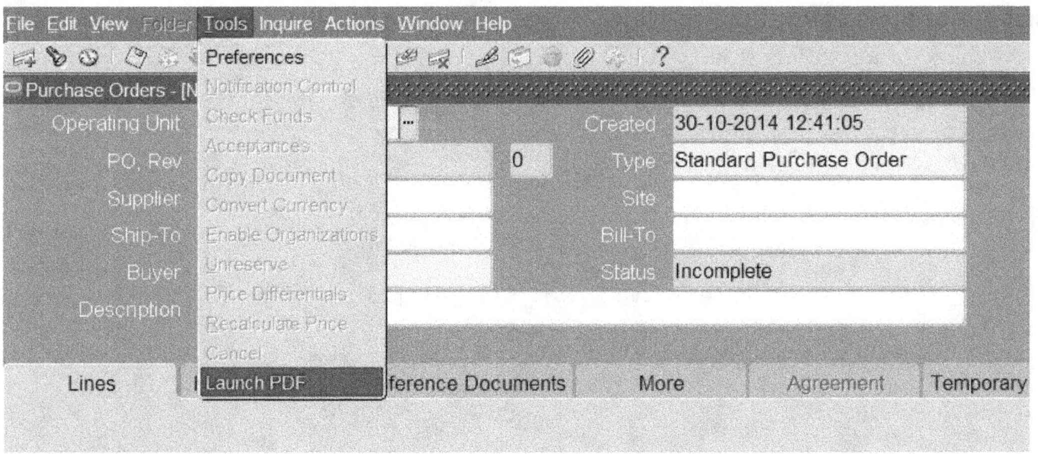

At this stage, nothing really will happen if you click on this submenu. The objective here is to launch the "PO Output for Communication" concurrent program to view the PDF. We need to perform another personalization to achieve this.

<u>Step-2:</u>

Enable "Launch PDF" submenu to launch the "PO Output for Communication" program.

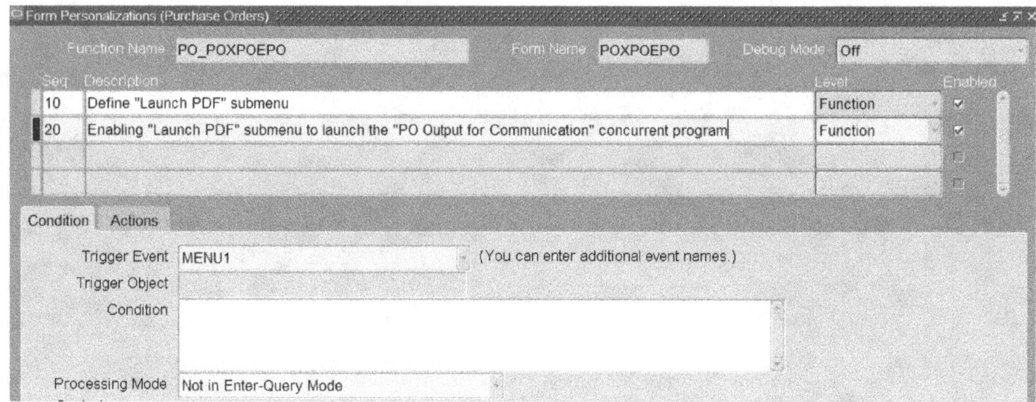

Now adding the sequence 20.

Seq: 20

Description: Enabling "Launch PDF" submenu to launch the "PO Output for Communication" concurrent program

Level: Function

Enabled: Yes

Condition:

Trigger Event: MENU1 (This value should be same as the one you had selected in the Menu Entry field of Seq-10 personalization. Refer to the screen shots above.)

Condition: <You may enter your condition here>

Processing Mode: Not in Enter-Query Mode.

Actions:

Seq: 10

Type: Builtin

Description: Launching the Program

Language: All

Enabled: Yes

Builtin Type: Launch SRS Form

Program Name: PO Output for Communication

Save the setup.

Now if you click on the "Launch PDF" submenu under the tools menu, you will notice the "Submit Request" form preloaded with "PO Output for Communication" program will appear as shown below.

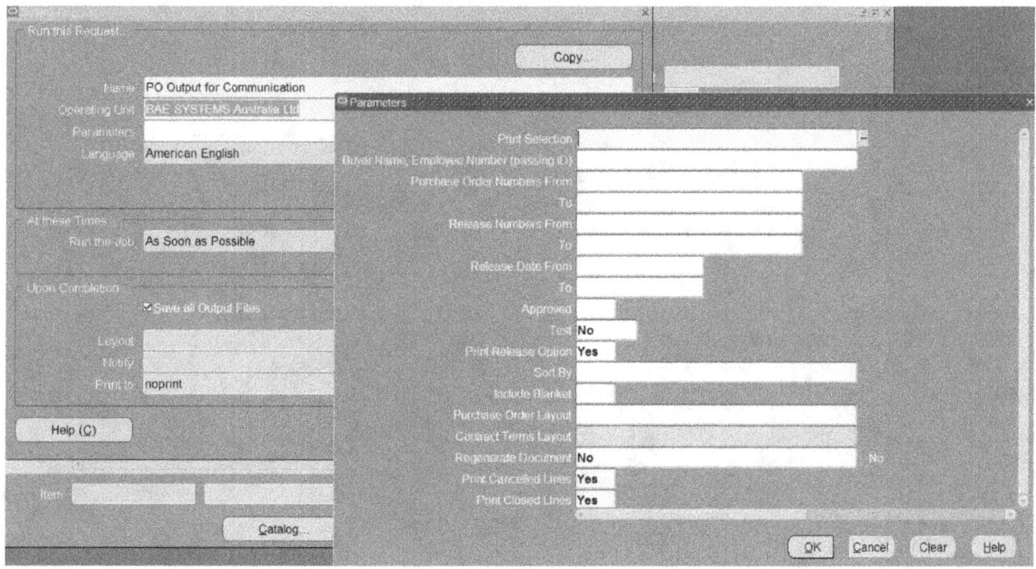

You need to select few parameter values and click on the "OK" button to submit the program.

This personalization will help you to save several mouse clicks to launch the concurrent program "PO Output for Communication".

The Builtin type "Launch SRS Form" launches the "Submit Request" form that is usually available through View -> Request and Submit a new request navigation.

Now we have seen how MENU and SPECIAL personalization can help us to establish an improved Oracle e-Businesss suite experience but you can further improve the user experience by auto populating the "Purchase Order Numbers From" and "To" fields in the above example. We will explain how to perform that magic in the subsequent chapters.

Object Type: Item

This chapter explains various Personalization properties available in the "Action" section of the Forms Personalization window. Each item has been explained based on real life business scenarios and you will be able to make Oracle Applications very user friendly once you learn each item and apply them in your area of business accordingly.

The "Item" property allows us to setup various physical properties of Oracle form fields, such as – Prompt, Visible, Read-Only, Enabled, Height, Width etc.

There are at least 50 properties available under the "Item" object type and we will discuss each object and property with valid business scenarios in the subsequent sections.

Before we get into the discussion of each object types and properties let me clarify the APPLICATION COVER and ITEM INSTANCE related properties.

You may come across personalization property that is post fixed with either APPLICATION COVER or ITEM-INSTANCE. For example, there are both DISPLAYED and DISPLAYED (APPLICATIONS COVER) as well as NAVIGABLE and NAVIGABLE (ITEM-INSTANCE) properties available for configuration.

APPLICATION COVER uses the Oracle e-Business suite routine APP_ITEM_PROPERTY.SET_PROPERTY which is the cover routine to the Oracle Forms built-in routine SET_ITEM_PROPERTY. The cover routine actually sets up multiple properties instead of just the targeted one. Please refer to the "Setting Properties" section of the Oracle e-Business Suite Developers guide to understand the key differences between SET_ITEM_PROPERTY and APP_ITEM_PROPERTY.SET_PROPERTY.

ITEM-INSTANCE is effective when you are dealing with a multi-record block. It actually sets the mentioned attribute for just one instance of the related item.

Now we will start discussing about each properties of the "Item" Object Type.

Property: ALTERABLE (APPLICATION COVER)

The ALTERABLE (APPLICATION COVER) property is used to allow or disallow INSERT and UPDATE activities on a specific instance of an item regardless of the activity (new record or update record) being performed on the row. You can still navigate through the item regardless of the property value setup (Yes/No).

Example:

Let's use the Purchase Order form to understand this property better.

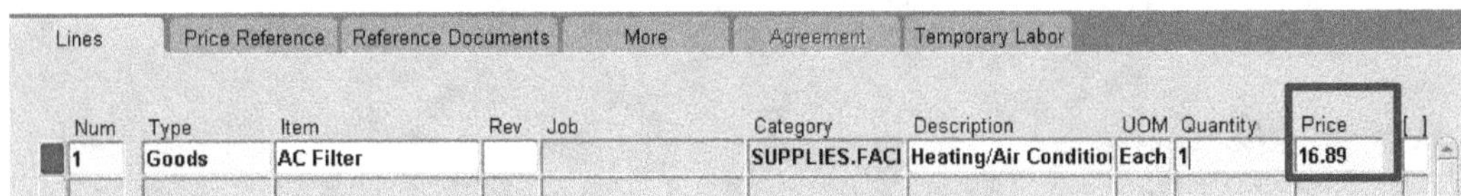

As per the above screen shot, the price field has been auto populated by Oracle based on the item entered into the PO Line level. But the business may require the price field to be non-editable to restrict users from updating the price.

In order to achieve the above requirement, you may use the ALTERABLE (APPLICATION COVER) property as below.

- Open the Forms Personalization window by clicking on the Help -> Diagnostics -> Custom Code -> Personalize menu option.
    - Seq – 10
    - Description – List Price read only
    - Level – Function
    - Enabled – Yes
- Condition:
    - Trigger Event - WHEN-NEW-ITEM-INSTANCE
    - Trigger Object - PO_LINES.UNIT_PRICE
    - Condition - :PO_LINES.UNIT_PRICE > 0
    - Processing Mode – Not in Enter-Query Mode
- Actions:
    - Seq – 10
    - Type – Property
    - Description - List Price Read-only
    - Language – All
    - Enabled – Yes
    - Object Type – Item
    - Target Object - PO_LINES.UNIT_PRICE
    - Property Name - ALTERABLE (APPLICATIONS COVER)
    - Value – False

Save the record.

Now if you close the form and reopen and try to enter a new PO line, you will notice, as soon as you enter into the price field after entering the quantity information, the price field will turn un-editable as shown below.

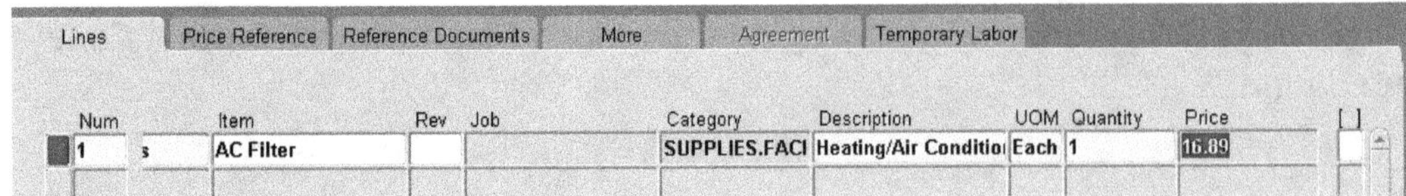

Note that, you can still navigate through the price field (as highlighted above) though it is un-editable but the same isn't possible when using other properties to fulfil the same requirement as discussed later in the ENTERABLE property discussion.

Property: ALTERABLE_PLUS and ALTERABLE_PLUS (APPLICATION COVER)

ALTERABLE_PLUS property is similar to the ALTERABLE property discussed above but only difference is – ALTERABLE covers a specific instance of an item; that means only the instance available in the current row; on the contrary the ALTERABLE_PLUS covered all instances of an item on a particular data block.

Significance of APPLICATION COVER has been already discussed in the previous sections.

Property: BACKGROUND_COLOR

As the name suggests you will be able to change the background color of various form fields using this property.

Let's try to highlight Purchase Order Number field in the Purchase Order form to demonstrate the personalization.

- Open the Forms Personalization window by clicking on the Help -> Diagnostics -> Custom Code -> Personalize menu option.
    - Seq – 10
    - Description – Background Color change
    - Level – Function
    - Enabled – Yes
- Condition:
    - Trigger Event - WHEN-NEW-FORM-INSTANCE
    - Processing Mode – Not in Enter-Query Mode
- Actions:
    - Seq – 10

- Type – Property
- Description – Background Color Change
- Language – All
- Enabled – Yes
- Object Type – Item
- Target Object - PO_HEADERS.SEGMENT1
- Property Name – BACKGROUND_COLOR
- Value – r0g255b0

Save the record.

As we might already know, RGB stands for Red, Green and Blue and maximum value can be assigned to each of them is 255; therefore r0g255b0 indicates "Green" color; similarly r0g0b255 indicates "red" and so on. You may experiment further with the values, such as r100g200b150 etc. and generate unique colors for the background.

Property: CASE_INSENSITIVE_QUERY

This property allows you to control the case sensitivity of a form field. For example – if you would like to search for a concurrent program in the concurrent program definition form accessible using the Application Developer responsibility through the Concurrent -> Program navigation, you don't need to mention the program name in the "Program" field maintaining proper case. The field is case-insensitive. On the other hand, if you want to search for the move orders using Inventory Superuser or similar responsibility mentioning the transaction type value, you need to be mindful of the case while typing the search condition.

Actually I have come across many scenarios, where users were reluctant to follow proper case while searching and we had to personalize respective field to make it case insensitive.

We will demonstrate this personalization using the transaction type field (highlighted below) of the Move Order form.

Responsibility: Inventory Super User or similar

Navigation: Move Orders -> Move Orders

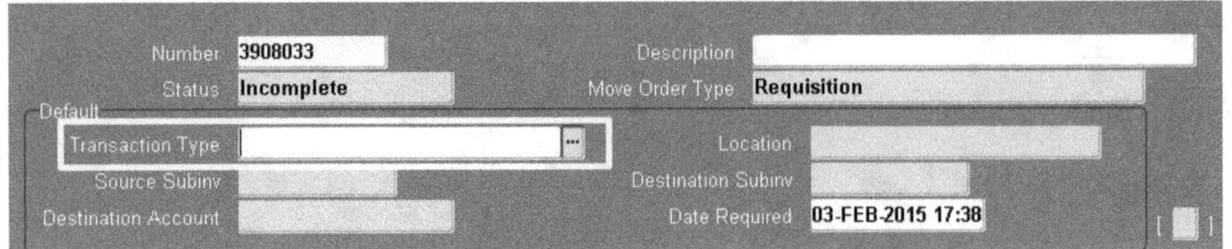

- Open the Forms Personalization window by clicking on the Help -> Diagnostics -> Custom Code -> Personalize menu option.
    - Seq – 10
    - Description – Move Order Transaction Type - Case Insensitivity setup
    - Level – Function
    - Enabled – Yes
- Condition:
    - Trigger Event - WHEN-NEW-FORM-INSTANCE
    - Processing Mode – Not in Enter-Query Mode
- Actions:
    - Seq – 10
    - Type – Property
    - Description – <Optional>
    - Language – All
    - Enabled – Yes
    - Object Type – Item
    - Target Object-TOMAI_MAIN_HEADER_BLK.TRANSACTION_TYPE_NAME
    - Property Name – CASE_INSENSITIVE_QUERY
    - Value – TRUE

Save the record.

Property: CASE_RESTRICTION

This property is used to setup UPPER or LOWER case for a form field. Regardless of the CAPS ON or OFF status in your keyboard, the personalization will be effected on the respective form field.

We will demonstrate this personalization using the Description field (visible in the above screen shot) of the Move Order form and try to make it an UPPERCASE field.

- Open the Forms Personalization window by clicking on the Help -> Diagnostics -> Custom Code -> Personalize menu option.
    - Seq – 10
    - Description – Move Order Header Description UPPERCASE setup
    - Level – Function
    - Enabled – Yes
- Condition:
    - Trigger Event - WHEN-NEW-FORM-INSTANCE
    - Processing Mode – Not in Enter-Query Mode
- Actions:
    - Seq – 10
    - Type – Property
    - Description – <Optional>
    - Language – All
    - Enabled – Yes
    - Object Type – Item
    - Target Object-TOMAI_MAIN_HEADER_BLK.DESCRIPTION
    - Property Name – CASE_RESTRICTION
    - Value – UPPERCASE (UPPERCASE/NONE/LOWERCASE)

Save the record.

Now, whatever you type in the description field it will be entered using UPPERCASE though CAPS lock in your keyboard is off.

Property: CONCEAL_DATA

This property will allow you to conceal the characters entered into a form field. Usually the confidential information like password, bank account number etc. needs to be concealed in any user interface.

We will demonstrate this personalization using the Description field of the Move Order form again.

- Open the Forms Personalization window by clicking on the Help -> Diagnostics -> Custom Code -> Personalize menu option.
  - Seq – 10
  - Description – Move Order Header Description CONCEAL DATA setup
  - Level – Function
  - Enabled – Yes
- Condition:
  - Trigger Event - WHEN-NEW-FORM-INSTANCE
  - Processing Mode – Not in Enter-Query Mode
- Actions:
  - Seq – 10
  - Type – Property
  - Description – <Optional>
  - Language – All
  - Enabled – Yes
  - Object Type – Item
  - Target Object-TOMAI_MAIN_HEADER_BLK.DESCRIPTION
  - Property Name – CONCEAL_DATA
  - Value – TRUE

Save the record.

Now the data in the description field will appear as below.

Description ****************

Property: DISPLAYED

As the name suggests, this property allows you to hide or display a form field/object based on the business requirements.

We will hide the Description field of the Move Order form to demonstrate this property.

- Open the Forms Personalization window by clicking on the Help -> Diagnostics -> Custom Code -> Personalize menu option.
  - Seq – 10
  - Description – Move Order Description DISPLAYED
  - Level – Function
  - Enabled – Yes
- Condition:
  - Trigger Event - WHEN-NEW-FORM-INSTANCE
  - Processing Mode – Not in Enter-Query Mode
- Actions:
  - Seq – 10
  - Type – Property
  - Description – <Optional>
  - Language – All
  - Enabled – Yes
  - Object Type – Item
  - Target Object-TOMAI_MAIN_HEADER_BLK.DESCRIPTION

- Property Name – DISPLAYED
- Value – TRUE

Save the record.

Property: ENABLED

ENABLED property allows you to enable or disable a form field/object based on the business requirements.

We will disable the Description field of the Move Order form to demonstrate this property.

- Open the Forms Personalization window by clicking on the Help -> Diagnostics -> Custom Code -> Personalize menu option.
    - Seq – 10
    - Description – Move Order Description field Disabled
    - Level – Function
    - Enabled – Yes
- Condition:
    - Trigger Event - WHEN-NEW-FORM-INSTANCE
    - Processing Mode – Not in Enter-Query Mode
- Actions:
    - Seq – 10
    - Type – Property
    - Description – <Optional>
    - Language – All
    - Enabled – Yes
    - Object Type – Item
    - Target Object-TOMAI_MAIN_HEADER_BLK.DESCRIPTION

- Property Name – ENABLED
- Value – FALSE

Save the record.

Property: ENTERABLE

ENTERABLE property controls the ability of a user to enter values in a form field/object. Referring to the ALTERABLE property discussed earlier in this chapter, ENTERABLE property performs similar to the ALTERABLE property but unlike the ALTERABLE property, you won't be able to navigate through a form field that has been personalized using ENTERABLE property.

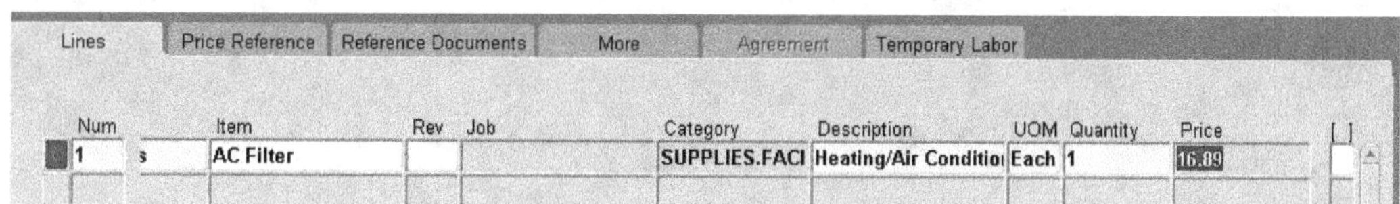

Just to explain the difference further, you can fulfil the requirement of making the "Price" field read-only as per the above screen shot using either ALTERABLE or ENTERABLE property but you won't be able to navigate through the price field (like it is highlighted in the above screen shot) if you using ENTERABLE property.

We will make the Description field of the Move Order form read-only to demonstrate this property.

- Open the Forms Personalization window by clicking on the Help -> Diagnostics -> Custom Code -> Personalize menu option.
  - Seq – 10
  - Description – Move Order Description field Read-Only
  - Level – Function
  - Enabled – Yes
- Condition:
  - Trigger Event - WHEN-NEW-FORM-INSTANCE
  - Processing Mode – Not in Enter-Query Mode
- Actions:
  - Seq – 10

- Type – Property
- Description – <Optional>
- Language – All
- Enabled – Yes
- Object Type – Item
- Target Object-TOMAI_MAIN_HEADER_BLK.DESCRIPTION
- Property Name – ENTERABLE
- Value – FALSE

Save the record.

Property: FOREGROUND_COLOR

This property sets up the text color of a form field. The configuration is similar to the Background Color property discussed earlier in this chapter.

We will make the Foreground Color of the Move Order Description field as "Blue" to demonstrate the functionality.

- Open the Forms Personalization window by clicking on the Help -> Diagnostics -> Custom Code -> Personalize menu option.
    - Seq – 10
    - Description – Move Order Description Foreground color change
    - Level – Function
    - Enabled – Yes
- Condition:
    - Trigger Event - WHEN-NEW-ITEM-INSTANCE
    - Trigger Object - TOMAI_MAIN_HEADER_BLK.DESCRIPTION
    - Processing Mode – Not in Enter-Query Mode
- Actions:

- Seq – 10
- Type – Property
- Description – Move Order Description Foreground Color
- Language – All
- Enabled – Yes
- Object Type – Item
- Target Object-TOMAI_MAIN_HEADER_BLK.DESCRIPTION
- Property Name – FOREGROUND_COLOR
- Value – r0g0b255

Save the record.

As mentioned earlier in this book, RGB stands for Red, Green and Blue and maximum value can be assigned to each of them is 255; therefore r0g0b255 indicates "Blue" color.

Property: FORMAT_MASK

If you need to fulfil a business requirement to enable the user to enter numeric values of a form field maintaining a specific pattern, such as 99,999,999.00; then you need to use the FORMAT_MASK form personalization.

We will demonstrate this functionality by modifying the formatting of the Move Order line quantity as 99,999,999.00 and that will automatically format the line quantity as 100,000,000.00 when we enter 100000000 in the quantity field.

- Open the Forms Personalization window by clicking on the Help -> Diagnostics -> Custom Code -> Personalize menu option.
    - Seq – 10
    - Description – Move Order line quantity FORMAT MASKING
    - Level – Function
    - Enabled – Yes
- Condition:

- Trigger Event - WHEN-NEW-ITEM-INSTANCE
- Trigger Object - TOMAI_MAIN_LINES_BLK.QUANTITY
- Processing Mode – Not in Enter-Query Mode

- Actions:
    - Seq – 10
    - Type – Property
    - Description – Format Mask 999,999,999.00
    - Language – All
    - Enabled – Yes
    - Object Type – Item
    - Target Object- TOMAI_MAIN_LINES_BLK.QUANTITY
    - Property Name – FORMAT_MASK
    - Value – FM99G999G999G999D00

Save the record.

Now the quantity value appears as below.

Transaction Type	Date Required	UOM	Primar Quantity
Move Order Issue	05-FEB-2015 19:01:5	Ea	100,000,000.00

Property: HEIGHT

As you expect, the HEIGHT property allow you to setup the height of a form field. Regular height for a text field is .25; therefore you may determine the planned value accordingly based on the business requirement. You need to be careful while increasing the height of a form field, as expanded form field may overlap the adjacent form fields and make them invisible.

We will try to explain this personalization by increasing the field size of the Item Description field in the Move Order form.

Before personalization the field looks as below.

- Open the Forms Personalization window by clicking on the Help -> Diagnostics -> Custom Code -> Personalize menu option.
    - Seq – 10
    - Description – Move Order Item Description height
    - Level – Function
    - Enabled – Yes
- Condition:
    - Trigger Event - WHEN-NEW-FORM-INSTANCE
    - Processing Mode – Not in Enter-Query Mode
- Actions:
    - Seq – 10
    - Type – Property
    - Description – Height Personalization
    - Language – All
    - Enabled – Yes
    - Object Type – Item
    - Target Object- TOMAI_MAIN_LINES_BLK.ITEM_DESCRIPTION
    - Property Name – HEIGHT
    - Value – .5

Save the record.

Now the item description field has expanded but it has partially overlapped with the adjacent fields as shown below.

Therefore it's important to assess the impact prior to increasing the height of a form field using this personalization.

Property: INITIAL_VALUE

This property allows setting up the initial value of a form field. You may use this personalization to setup a numeric, text or a date field.

We will setup up an initial value for the Date Required field in the Move Order header level to explain this personalization.

Please note that, the INITIAL_VALUE personalization can only be setup in the WHEN-NEW-RECORD-INSTANCE trigger.

Let's assume, based on the organization mandates, move orders are created 5 days prior to the actual requirement date; therefore you have been asked to develop a mechanism to default the date required field accordingly. Following steps will help us to deliver the requirement.

- Open the Forms Personalization window by clicking on the Help -> Diagnostics -> Custom Code -> Personalize menu option.
    - Seq – 10
    - Description – Move Order Date Required - Initial Value
    - Level – Function
    - Enabled – Yes
- Condition:
    - Trigger Event - WHEN-NEW-RECORD-INSTANCE
    - Trigger Object - TOMAI_MAIN_HEADER_BLK
    - Processing Mode – Not in Enter-Query Mode
- Actions:
    - Seq – 10
    - Type – Property

- Description – Move Order Date Required - Initial Value
- Language – All
- Enabled – Yes
- Object Type – Item
- Target Object- TOMAI_MAIN_HEADER_BLK.DATE_REQUIRED
- Property Name – INITIAL_VALUE
- Value – =SELECT TO_CHAR(SYSDATE+5,'DD-MON-YYYY HH24:MI:SS') FROM dual

Save the record.

This personalization also explains how to use a SQL statement to dynamically derive a value.

Property: INSERT_ALLOWED

This property controls the ability to insert values into a form field and behaves similar to ENTERABLE and ALTERABLE fields explained earlier. Setting up this property as FALSE surely restricts insertion but doesn't grey out the property.

We will further extend the example used above and assume that the business wants the Date Required field at the Move Order header level to be Read Only to enforce 5 days material issue lead time for all the move orders. Therefore in addition to defaulting SYSDATE+5 date time in the Date Required field (explained above), we also need to make it read-only to enforce the restriction.

Following personalization steps will be able to fulfil the requirement.

- Open the Forms Personalization window by clicking on the Help -> Diagnostics -> Custom Code -> Personalize menu option.
    - Seq – 10
    - Description – Move Order Date Required - INSERT ALLOWED
    - Level – Function
    - Enabled – Yes
- Condition:
    - Trigger Event - WHEN-NEW-RECORD-INSTANCE

- Trigger Object - TOMAI_MAIN_HEADER_BLK
- Processing Mode – Not in Enter-Query Mode

- Actions:
  - Seq – 10
  - Type – Property
  - Description – <OPTIONAL>
  - Language – All
  - Enabled – Yes
  - Object Type – Item
  - Target Object- TOMAI_MAIN_HEADER_BLK.DATE_REQUIRED
  - Property Name – INSERT_ALLOWED
  - Value – FALSE

Save the record.

Property: LABEL

This property is used to modify the label/caption of a form attribute. This is applicable only to the properties that have a label attached to it (e.g. buttons).

Let's assume you have been asked to modify the caption of the "Approve" button in the Move Orders form as "Manager Approval".

Following personalization steps will be able to fulfil the requirement.

- Open the Forms Personalization window by clicking on the Help -> Diagnostics -> Custom Code -> Personalize menu option.
  - Seq – 10
  - Description – Modify caption of the Approve button
  - Level – Function
  - Enabled – Yes

- Condition:
    - Trigger Event - WHEN-NEW-FORM-INSTANCE
    - Processing Mode – Not in Enter-Query Mode
- Actions:
    - Seq – 10
    - Type – Property
    - Description – <OPTIONAL>
    - Language – All
    - Enabled – Yes
    - Object Type – Item
    - Target Object- TOMAI_MAIN_HEADER_BLK.DONE
    - Property Name – LABEL
    - Value – Manager Approval

Save the record.

Property: NAVIGABLE

This property controls the navigable attribute of a form field; that means when you are using keyboard navigation, you won't be able to navigate to the field that has NAVIGABLE = False and vice versa.

For example, when you use keyboard navigation on the requisition line, tabbing out of the "Num" field will take you to the "Type" field if the NAVIGABLE property has been set as TRUE on the "Type" field; otherwise it will skip the "Type" field and navigate directly to the "Item" field.

Num	Type	Item	Rev	Category	Description
1	Goods				

Following personalization steps will make the "Type" field non-navigable.

- Open the Forms Personalization window by clicking on the Help -> Diagnostics -> Custom Code -> Personalize menu option.

- Seq – 10
- Description – <Enter meaningful Description>
- Level – Function
- Enabled – Yes

- Condition:
  - Trigger Event - WHEN-NEW-FORM-INSTANCE
  - Processing Mode – Not in Enter-Query Mode
- Actions:
  - Seq – 10
  - Type – Property
  - Description – <OPTIONAL>
  - Language – All
  - Enabled – Yes
  - Object Type – Item
  - Target Object- LINES.LINES_TYPE
  - Property Name – NAVIGABLE
  - Value – FALSE

Save the record.

Property: NEXT_NAVIGATION_ITEM

This property allows to directly navigating to the preferred form field, skipping one/many form fields. Appropriate set-up of navigation sequence through this personalization attribute may potentially save significant amount of data entry timing in various Oracle Forms.

Let us use the requisition form again to explain this personalization property and we will setup a personalization to directly navigate to the "Description" field from the "Type" field by skipping 3 forms fields as per below screen shot.

Num	Type	Item	Rev	Category	Description
1	Goods				

- Open the Forms Personalization window by clicking on the Help -> Diagnostics -> Custom Code -> Personalize menu option.
  - Seq – 10
  - Description – <Enter meaningful Description>
  - Level – Function
  - Enabled – Yes
- Condition:
  - Trigger Event - WHEN-NEW-FORM-INSTANCE
  - Processing Mode – Not in Enter-Query Mode
- Actions:
  - Seq – 10
  - Type – Property
  - Description – <OPTIONAL>
  - Language – All
  - Enabled – Yes
  - Object Type – Item
  - Target Object- LINES.LINES_TYPE
  - Property Name – NEXT_NAVIGATION_ITEM
  - Value – ITEM_DESCRIPTION

Save the record.

Property: PREVIOUS_NAVIGATION_ITEM

This attribute works when you use ALT+TAB key combination to navigate to the previous form field from the current form field unlike the NEXT_NAVIGATION_ITEM which works when you use TAB key to navigate to the next field.

Referring to the requisition form, as per the below screen shot, if you use the ALT+TAB key combination from the "Description" field, usually the navigation goes to the "Category" field. Now, let's setup a forms personalization to navigate directly back to "Num" field.

Num	Type	Item	Rev	Category	Description
1	Goods			110.33	V-Belts 10mm

- Open the Forms Personalization window by clicking on the Help -> Diagnostics -> Custom Code -> Personalize menu option.
    - Seq – 10
    - Description – <Enter meaningful Description>
    - Level – Function
    - Enabled – Yes
- Condition:
    - Trigger Event - WHEN-NEW-FORM-INSTANCE
    - Processing Mode – Not in Enter-Query Mode
- Actions:
    - Seq – 10
    - Type – Property
    - Description – <OPTIONAL>
    - Language – All
    - Enabled – Yes
    - Object Type – Item
    - Target Object- LINES.ITEM_DESCRIPTION

- Property Name – PREVIOUS_NAVIGATION_ITEM
- Value – LINE_NUM

Save the record.

Property: PROMPT_ALIGNMENT_OFFSET

This property helps us to set alignment for a form field prompt. Below screen shots explain the utility of this personalization.

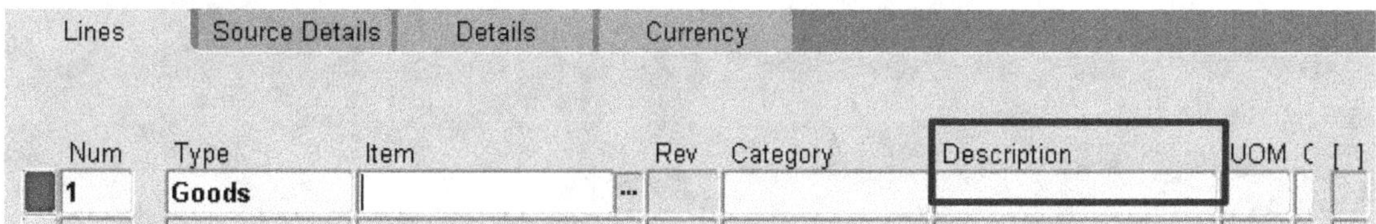

As per the above screen shot, the prompt "Description" is left aligned with the field. Now if we want to align it little towards Centre or right, then following personalization will help us to achieve the same.

- Open the Forms Personalization window by clicking on the Help -> Diagnostics -> Custom Code -> Personalize menu option.
    - Seq – 10
    - Description – <Enter meaningful Description>
    - Level – Function
    - Enabled – Yes
- Condition:
    - Trigger Event - WHEN-NEW-FORM-INSTANCE
    - Processing Mode – Not in Enter-Query Mode
- Actions:
    - Seq – 10
    - Type – Property
    - Description – <OPTIONAL>
    - Language – All

- Enabled – Yes
- Object Type – Item
- Target Object- LINES.ITEM_DESCRIPTION
- Property Name – PROMPT_ALIGNMENT_OFFSET
- Value – 0.5 (initial value was 0.05)

Save the record.

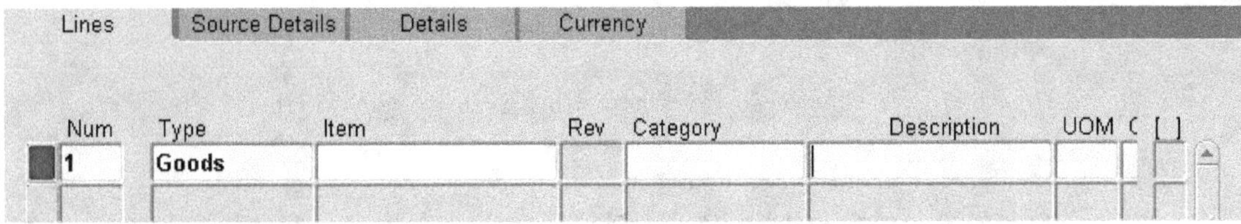

The "Description" field is aligned differently now. You can adjust the value to fine tune the alignment further.

Property: PROMPT_EDGE_OFFSET

This property enabled us to setup the vertical offsetting of the form field prompt. Following screen shots and configurations will help you to understand this property better.

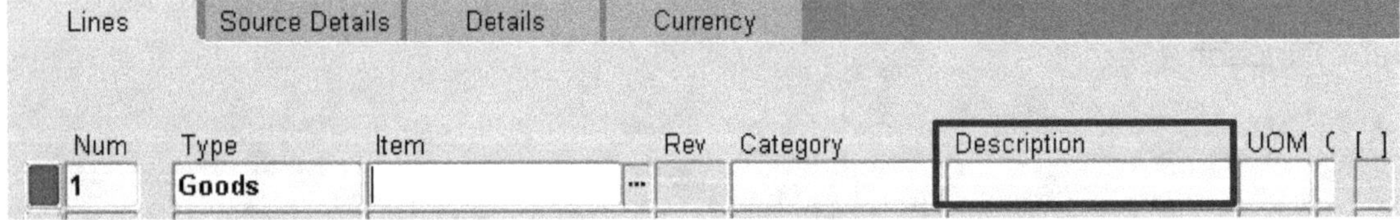

Referring to the requisition line screen shot above, we can see that there is no-gap between the form field and the prompt. If you want to introduce little gap between the prompt and the form field then below configuration will help.

- Open the Forms Personalization window by clicking on the Help -> Diagnostics -> Custom Code -> Personalize menu option.
    - Seq – 10
    - Description – <Enter meaningful Description>
    - Level – Function
    - Enabled – Yes

- Condition:
  - Trigger Event - WHEN-NEW-FORM-INSTANCE
  - Processing Mode – Not in Enter-Query Mode
- Actions:
  - Seq – 10
  - Type – Property
  - Description – <OPTIONAL>
  - Language – All
  - Enabled – Yes
  - Object Type – Item
  - Target Object- LINES.ITEM_DESCRIPTION
  - Property Name – PROMPT_EDGE_OFFSET
  - Value – 0.2 (initial value was 0)

Save the record.

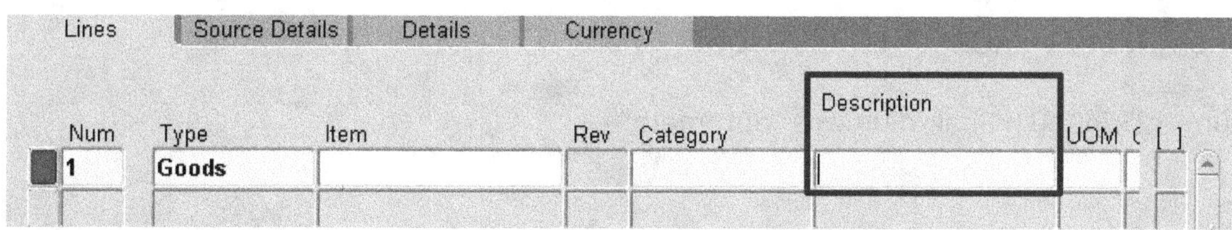

Above screen shot shows the difference.

Property: PROMPT_FOREGROUND_COLOR

As the name suggests, this property allows us to setup the foreground color of the form fields. We will demonstrate this personalization by changing the foreground color of the "Description" prompt to Red based on the below personalization steps.

- Open the Forms Personalization window by clicking on the Help -> Diagnostics -> Custom Code -> Personalize menu option.
  - Seq – 10

- Description – <Enter meaningful Description>
- Level – Function
- Enabled – Yes

- Condition:
  - Trigger Event - WHEN-NEW-FORM-INSTANCE
  - Processing Mode – Not in Enter-Query Mode
- Actions:
  - Seq – 10
  - Type – Property
  - Description – <OPTIONAL>
  - Language – All
  - Enabled – Yes
  - Object Type – Item
  - Target Object- LINES.ITEM_DESCRIPTION
  - Property Name – PROMPT_FOREGROUND_COLOR
  - Value – r255g0b0 (initial value was "automatic")

Save the record.

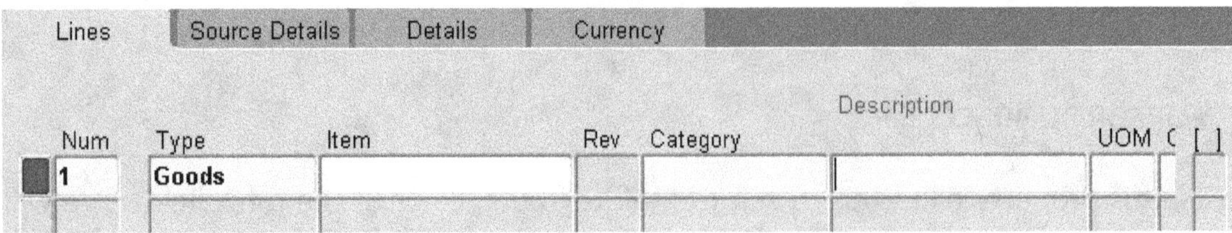

Now the "Description" prompt has turned red.

The significance of the color coding "r255g0b0' has been already explained earlier in this book.

Property: PROMPT_TEXT

This is a very frequently used personalization property that is used to update the form field prompt to make it more user friendly based on the business requirements. We will demonstrate this property by changing the "Description" prompt to "Item Description".

- Open the Forms Personalization window by clicking on the Help -> Diagnostics -> Custom Code -> Personalize menu option.
    - Seq – 10
    - Description – <Enter meaningful Description>
    - Level – Function
    - Enabled – Yes
- Condition:
    - Trigger Event - WHEN-NEW-FORM-INSTANCE
    - Processing Mode – Not in Enter-Query Mode
- Actions:
    - Seq – 10
    - Type – Property
    - Description – <OPTIONAL>
    - Language – All
    - Enabled – Yes
    - Object Type – Item
    - Target Object- LINES.ITEM_DESCRIPTION
    - Property Name – PROMPT_TEXT
    - Value – Item Description (initial value was "Description")

Save the record.

Above screen shot shows the change.

Property: QUERYABLE

This property controls the ability to query record based on a particular form field value. For example: it is allowed to query the requisition details using the requisition number in the below window.

The above screen shot indicates that, the form is now in query mode and we are trying to query requisition number – 14324.

Your business may ask to remove the ability for the users to query requisitions based on requisition number in requisition form while encouraging users to use the Requisition Summary form instead.

Following personalization steps will help you to fulfil the requirement.

- Open the Forms Personalization window by clicking on the Help -> Diagnostics -> Custom Code -> Personalize menu option.
    - Seq – 10
    - Description – <Enter meaningful Description>
    - Level – Function
    - Enabled – Yes
- Condition:
    - Trigger Event - WHEN-NEW-FORM-INSTANCE
    - Processing Mode – Not in Enter-Query Mode
- Actions:
    - Seq – 10

- Type – Property
- Description – <OPTIONAL>
- Language – All
- Enabled – Yes
- Object Type – Item
- Target Object- PO_REQ_HDR.SEGMENT1
- Property Name – QUERYABLE
- Value – TRUE (initial value was “FALSE”)

Save the record.

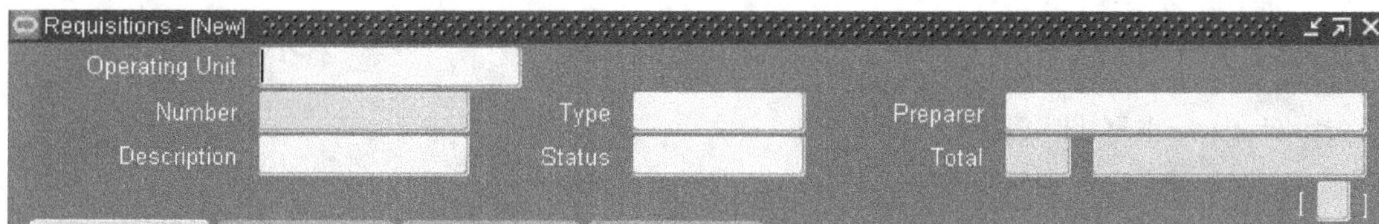

Above screen shot indicates that, the requisition number field is greyed out in the query-mode; therefore user won’t be able to enter the requisition number to query a requisition in the requisition form.

Property: REQUIRED

You can make a field mandatory by using this personalization property. We will make the Need-by date field mandatory in the requisition lines block to demonstrate this property.

- Open the Forms Personalization window by clicking on the Help -> Diagnostics -> Custom Code -> Personalize menu option.
    - Seq – 10
    - Description – <Enter meaningful Description>
    - Level – Function
    - Enabled – Yes
- Condition:
    - Trigger Event - WHEN-NEW-BLOCK-INSTANCE

- Trigger Object - LINES
- Processing Mode – Not in Enter-Query Mode

- Actions:
    - Seq – 10
    - Type – Property
    - Description – <OPTIONAL>
    - Language – All
    - Enabled – Yes
    - Object Type – Item
    - Target Object- LINES.NEED_BY_DATE
    - Property Name – REQUIRED
    - Value – TRUE

Save the record.

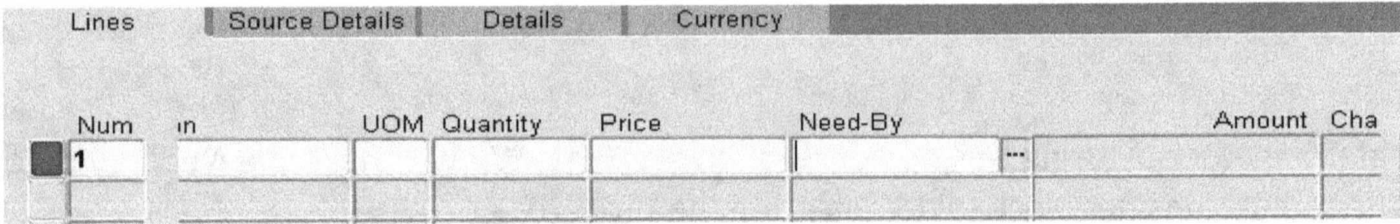

Above screen shot shows that the "Need-By" date field is now mandatory.

Property: TOOLTIP_BACKGROUND_COLOR

Before we discuss about this property, we need to note that, not all the form field shows the tooltips. This property allows you to change the background of the tooltips wherever it appears. We will use the Routing Resource definition form in the Bill of Materials responsibility to demonstrate this property.

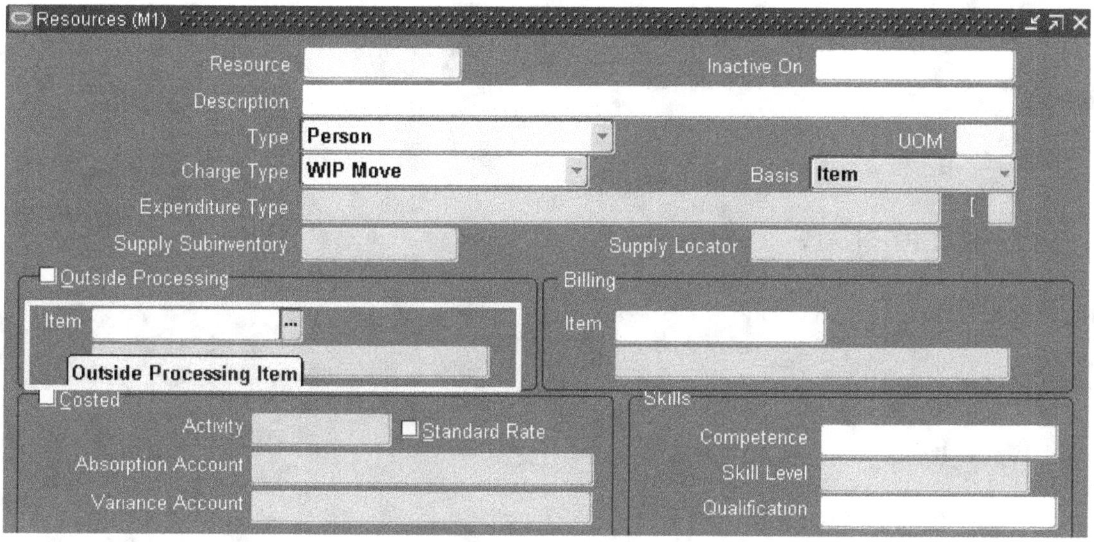

Above form is accessible through below responsibility and navigation:

Responsibility: Bill of Materials or similar

Navigation: Routing -> Resources.

Now, if you place your mouse just in the "Item" field in the "Outside Processing" section, it will show the tooltips as shown above.

We will change the background color of the above tooltips from "Yellow" to "Blue" using TOOLTIP_BACKGROUND_COLOR property as shown below.

- Open the Forms Personalization window by clicking on the Help -> Diagnostics -> Custom Code -> Personalize menu option.
    - Seq – 10
    - Description – <Enter meaningful Description>
    - Level – Function
    - Enabled – Yes
- Condition:
    - Trigger Event - WHEN-NEW-ITEM-INSTANCE
    - Trigger Object – B_RESOURCES.ITEM_FLEX_VALUES
    - Processing Mode – Not in Enter-Query Mode

- Actions:
    - Seq – 10
    - Type – Property
    - Description – <OPTIONAL>
    - Language – All
    - Enabled – Yes
    - Object Type – Item
    - Target Object- B_RESOURCES.ITEM_FLEX_VALUES
    - Property Name: TOOLTIP_BACKGROUND_COLOR
    - Value – r0g0b255

Save the record.

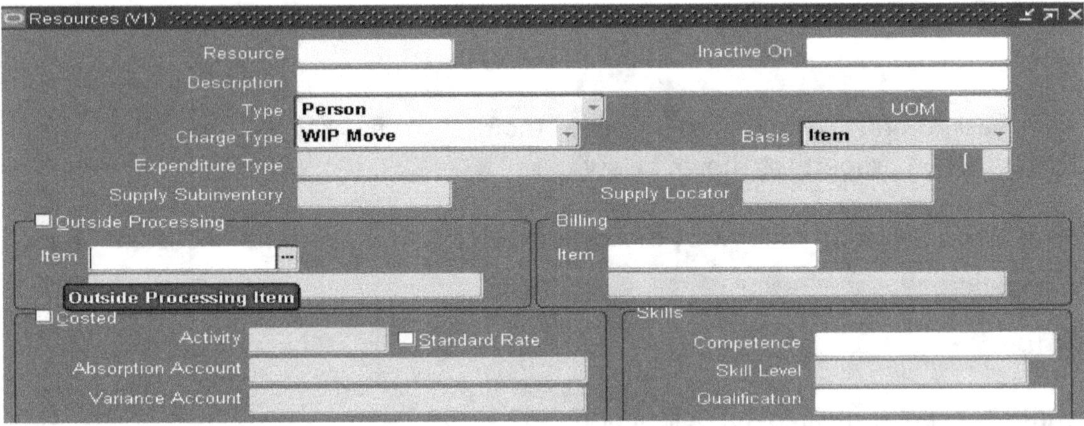

Now the tooltips appear in "Blue".

Property: TOOLTIP_FOREGROUND_COLOR

We will try to change the foreground color of the below tooltips from white to Green.

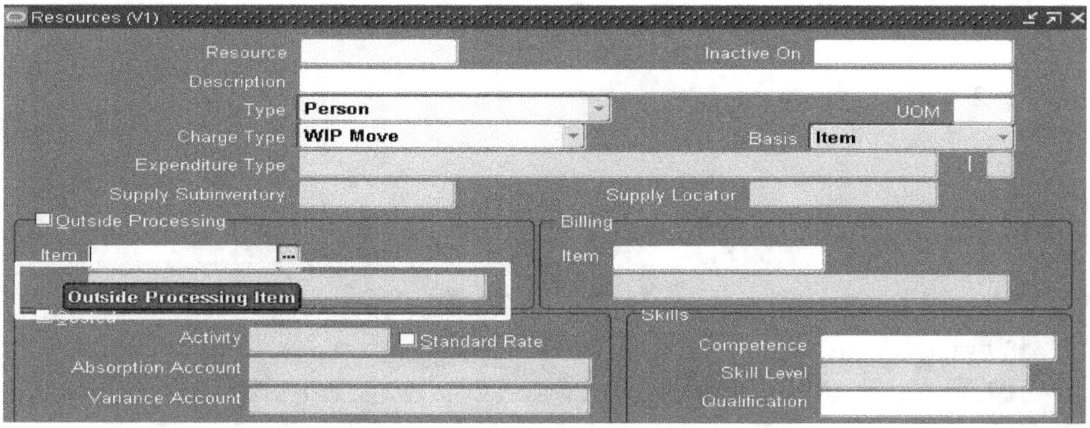

- Open the Forms Personalization window by clicking on the Help -> Diagnostics -> Custom Code -> Personalize menu option.
    - Seq – 10
    - Description – <Enter meaningful Description>
    - Level – Function
    - Enabled – Yes
- Condition:
    - Trigger Event - WHEN-NEW-ITEM-INSTANCE
    - Trigger Object – B_RESOURCES.ITEM_FLEX_VALUES
    - Processing Mode – Not in Enter-Query Mode
- Actions:
    - Seq – 10
    - Type – Property
    - Description – <OPTIONAL>
    - Language – All
    - Enabled – Yes
    - Object Type – Item
    - Target Object- B_RESOURCES.ITEM_FLEX_VALUES

- Property Name: TOOLTIP_FOREGROUND_COLOR
- Value – r0g255b0

Save the record.

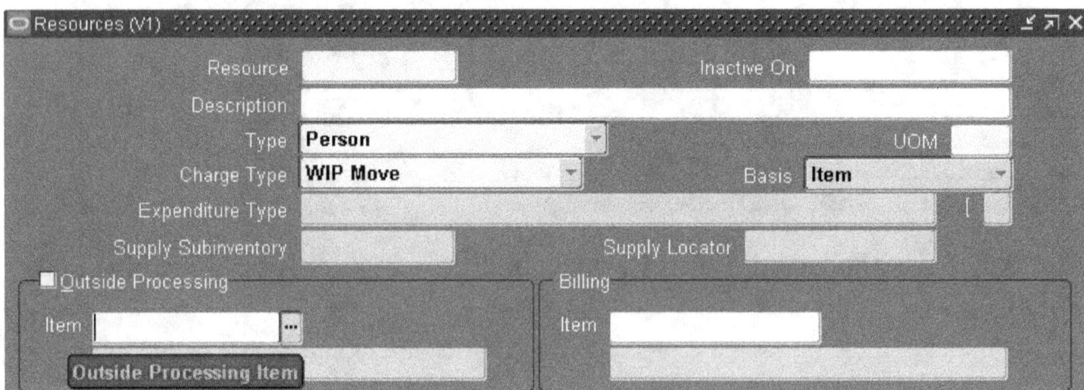

The tooltips now appears in Green.

Property: TOOLTIP_TEXT

This property allows us to change the tooltip text and we will change the above tooltip text to "OSP Item".

- Open the Forms Personalization window by clicking on the Help -> Diagnostics -> Custom Code -> Personalize menu option.
    - Seq – 10
    - Description – <Enter meaningful Description>
    - Level – Function
    - Enabled – Yes
- Condition:
    - Trigger Event - WHEN-NEW-ITEM-INSTANCE
    - Trigger Object – B_RESOURCES.ITEM_FLEX_VALUES
    - Processing Mode – Not in Enter-Query Mode
- Actions:
    - Seq – 10

- Type – Property
- Description – <OPTIONAL>
- Language – All
- Enabled – Yes
- Object Type – Item
- Target Object- B_RESOURCES.ITEM_FLEX_VALUES
- Property Name: TOOLTIP_TEXT
- Value – OSP Item

Save the record.

Tooltip text change is shown above.

Property: VALUE

This property allows you to set up a default value for a form field. We will set up a default value for the description field in the requisition header block to demonstrate the functionality.

- Open the Forms Personalization window by clicking on the Help -> Diagnostics -> Custom Code -> Personalize menu option.
  - Seq – 10
  - Description – <Enter meaningful Description>
  - Level – Function
  - Enabled – Yes

- Condition:
    - Trigger Event - WHEN-NEW-FORM-INSTANCE
    - Processing Mode – Not in Enter-Query Mode
- Actions:
    - Seq – 10
    - Type – Property
    - Description – <OPTIONAL>
    - Language – All
    - Enabled – Yes
    - Object Type – Item
    - Target Object- PO_REQ_HDR.DESCRIPTION
    - Property Name: VALUE
    - Value – Test Description

Save the record.

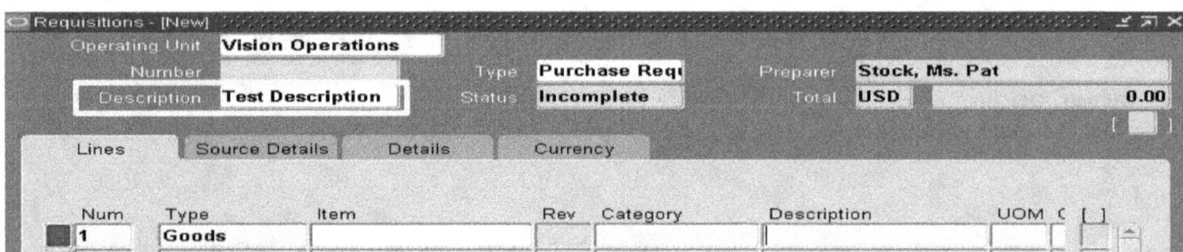

Above screen shot shows the default value.

Property: WIDTH

This property allows you setup the width of a form field and this is very useful when it's difficult for users to see the full content of a form field due to the smaller width.

We will expand the type field in the requisition header block to demonstrate the property, as based on the below screen shot, it's difficult to see the full text of the requisition type.

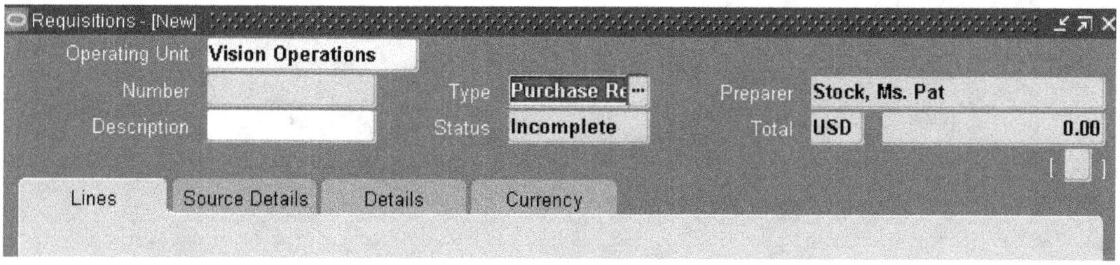

- Open the Forms Personalization window by clicking on the Help -> Diagnostics -> Custom Code -> Personalize menu option.
  - Seq – 10
  - Description – <Enter meaningful Description>
  - Level – Function
  - Enabled – Yes
- Condition:
  - Trigger Event - WHEN-NEW-FORM-INSTANCE
  - Processing Mode – Not in Enter-Query Mode
- Actions:
  - Seq – 10
  - Type – Property
  - Description – <OPTIONAL>
  - Language – All
  - Enabled – Yes
  - Object Type – Item
  - Target Object- PO_REQ_HDR.DOCUMENT_TYPE_DISPLAY
  - Property Name: WIDTH
  - Value – 2 (Initial value was "1")

Save the record.

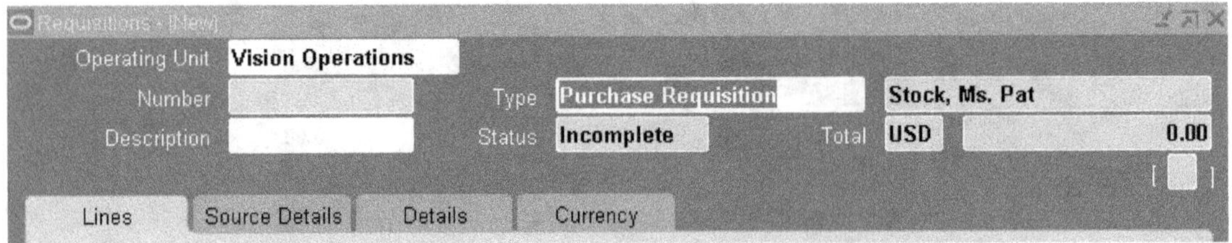

Now you can see the full text of the requisition type as shown in the above screen shot.

Property: X_POS

This property allows you to set up the horizontal position of a form field. We will move the “Type” field in the requisition header block towards right direction to explain this property.

You can see the initial horizontal location of the “Type” field in the above screen shot.

- Open the Forms Personalization window by clicking on the Help -> Diagnostics -> Custom Code -> Personalize menu option.
  - Seq – 10
  - Description – <Enter meaningful Description>
  - Level – Function
  - Enabled – Yes
- Condition:
  - Trigger Event - WHEN-NEW-FORM-INSTANCE
  - Processing Mode – Not in Enter-Query Mode
- Actions:
  - Seq – 10
  - Type – Property
  - Description – <OPTIONAL>
  - Language – All
  - Enabled – Yes
  - Object Type – Item

- Target Object- PO_REQ_HDR.DOCUMENT_TYPE_DISPLAY
- Property Name: X_POS
- Value – 4 (Initial value was 3.5)

Save the record.

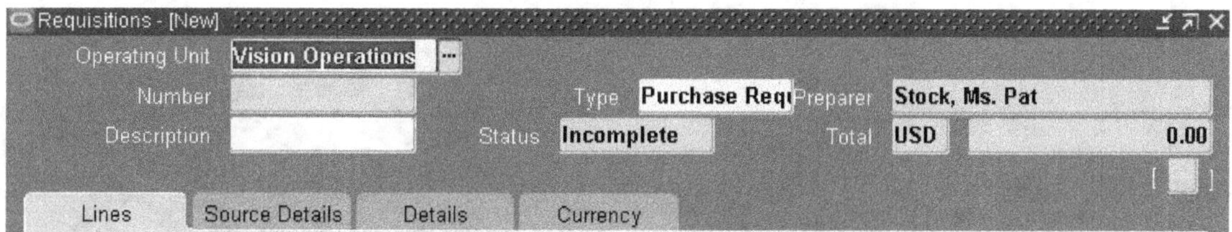

Now you can see the new horizontal location of the property.

Property: Y_POS

This property allows you to set up the vertical position of a form field. We will move the "Type" field in the requisition header block little upwards to explain this property.

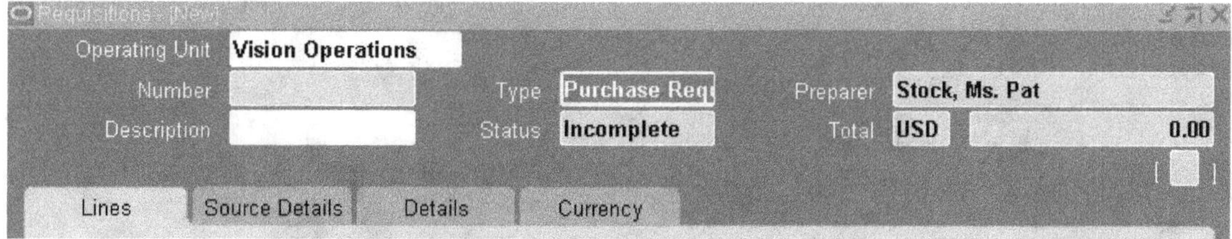

You can see the initial vertical location of the "Type" field in the above screen shot.

- Open the Forms Personalization window by clicking on the Help -> Diagnostics -> Custom Code -> Personalize menu option.
    - Seq – 10
    - Description – <Enter meaningful Description>
    - Level – Function
    - Enabled – Yes
- Condition:
    - Trigger Event - WHEN-NEW-FORM-INSTANCE
    - Processing Mode – Not in Enter-Query Mode

- Actions:
    - Seq – 10
    - Type – Property
    - Description – <OPTIONAL>
    - Language – All
    - Enabled – Yes
    - Object Type – Item
    - Target Object- PO_REQ_HDR.DOCUMENT_TYPE_DISPLAY
    - Property Name: Y_POS
    - Value – 0.1 (Initial value was .25)

Save the record.

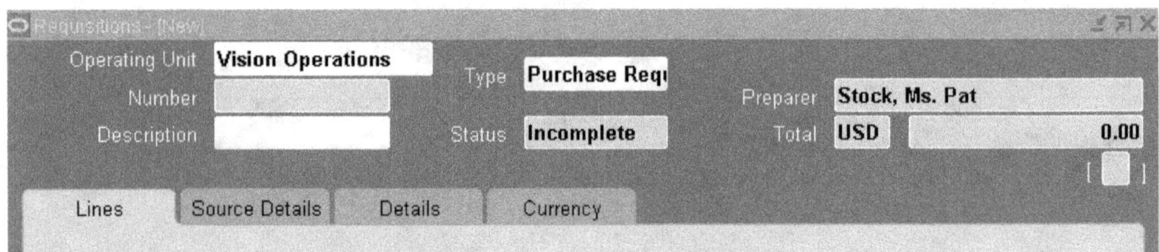

Now you can see the new vertical location of the property.

Chapter

4

Object Type: Window

This chapter explains various Personalization properties available for the object type "Window". You will be able to change various Window attributes once you learn the topics discussed in this chapter.

Property: TITLE

This property enables you to change the title of a window. We will change the title of the "Users" window accessible through Security -> User -> Define navigation in the "System Administrator" responsibility.

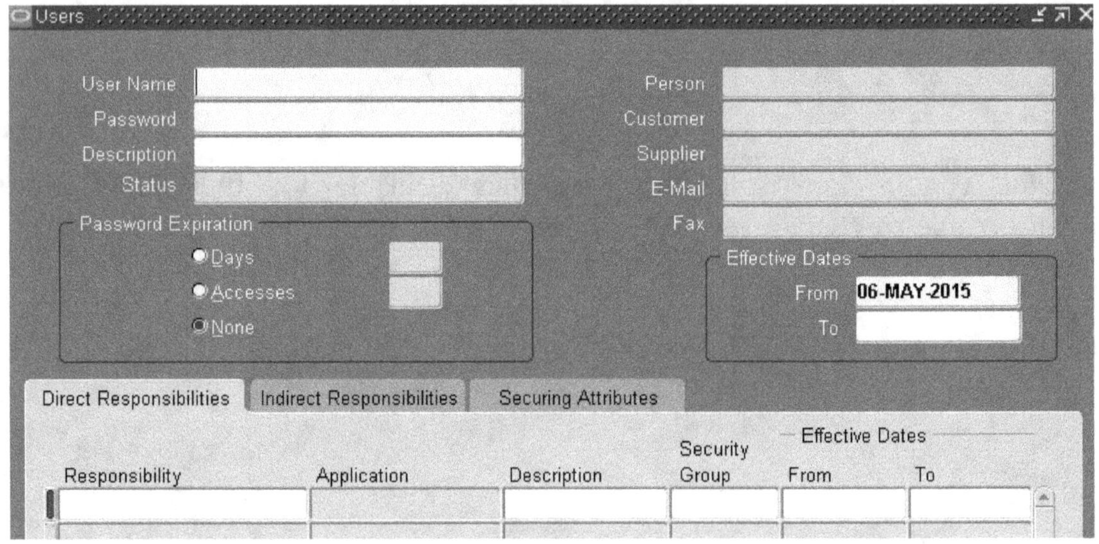

Above screen shot shows the initial title "Users". We will now change it to "Users of OraAppsGuide.com" using the "Title" property.

- Open the Forms Personalization window by clicking on the Help -> Diagnostics -> Custom Code -> Personalize menu option.
    - Seq – 10
    - Description – <Enter meaningful Description>
    - Level – Function
    - Enabled – Yes
- Condition:
    - Trigger Event - WHEN-NEW-FORM-INSTANCE
    - Processing Mode – Not in Enter-Query Mode
- Actions:
    - Seq – 10

- Type – Property
- Description – <OPTIONAL>
- Language – All
- Enabled – Yes
- Object Type – Window
- Target Object- USER_WINDOW
- Property Name: TITLE
- Value – Users of OraAppsGuide.com

Save the record.

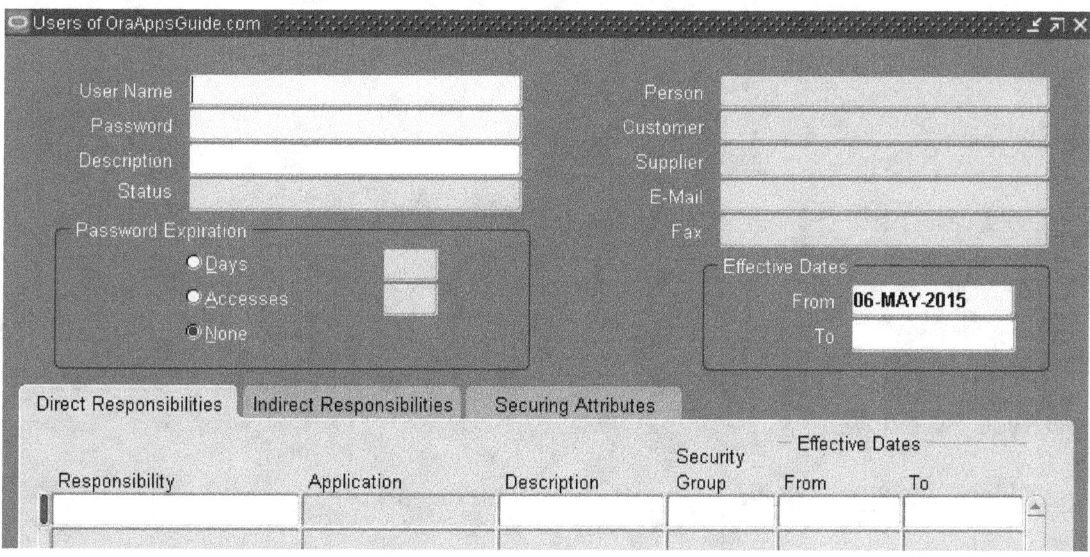

Above screen shot shows the changed title.

Property: HEIGHT

This property enables you to change the height of a window. We will increase the height of the "Users" window accessible through Security -> User -> Define navigation in the "System Administrator" responsibility.

This is useful when your form has several attributes and we need to scroll down to see all of them. You can just increase the height instead, so that all the attributes appear in a single window view.

- Open the Forms Personalization window by clicking on the Help -> Diagnostics -> Custom Code -> Personalize menu option.
    - Seq – 10
    - Description – <Enter meaningful Description>
    - Level – Function
    - Enabled – Yes
- Condition:
    - Trigger Event - WHEN-NEW-FORM-INSTANCE
    - Processing Mode – Not in Enter-Query Mode
- Actions:
    - Seq – 10
    - Type – Property
    - Description – <OPTIONAL>
    - Language – All
    - Enabled – Yes
    - Object Type – Window
    - Target Object- USER_WINDOW
    - Property Name: HEIGHT
    - Value – 7 (Initial value was 5)

Save the record.

Property: WIDTH

This property enables you to change the width of a window. We will increase the width of the "Users" window accessible through Security -> User -> Define navigation in the "System Administrator" responsibility.

This is useful when your form has several attributes and we need to scroll right to see all of them. You can just increase the width instead, so that all the attributes appear in a single window view.

- Open the Forms Personalization window by clicking on the Help -> Diagnostics -> Custom Code -> Personalize menu option.
    - Seq – 10
    - Description – <Enter meaningful Description>
    - Level – Function
    - Enabled – Yes
- Condition:
    - Trigger Event - WHEN-NEW-FORM-INSTANCE
    - Processing Mode – Not in Enter-Query Mode
- Actions:
    - Seq – 10
    - Type – Property
    - Description – <OPTIONAL>
    - Language – All
    - Enabled – Yes
    - Object Type – Window
    - Target Object- USER_WINDOW
    - Property Name: WIDTH
    - Value – 10 (Initial value was 7.802)

Save the record.

Property: WINDOW_STATE

This property enables you to setup the default state of a window when it opens, i.e. Maximize, Minimize or Normal.

Many organizations open the form windows in a minimized state if the form contains confidential information, such as HR and Payroll information.

We will again use the "Users" form to demonstrate this property and open this form in a minimized mode upon clicking on the menu option.

- Open the Forms Personalization window by clicking on the Help -> Diagnostics -> Custom Code -> Personalize menu option.
    - Seq – 10
    - Description – <Enter meaningful Description>
    - Level – Function
    - Enabled – Yes
- Condition:
    - Trigger Event - WHEN-NEW-FORM-INSTANCE
    - Processing Mode – Not in Enter-Query Mode
- Actions:
    - Seq – 10
    - Type – Property
    - Description – <OPTIONAL>
    - Language – All
    - Enabled – Yes
    - Object Type – Window
    - Target Object- USER_WINDOW
    - Property Name: WINDOW_STATE
    - Value – MINIMIZE (Initial value was 7.802)

Save the record.

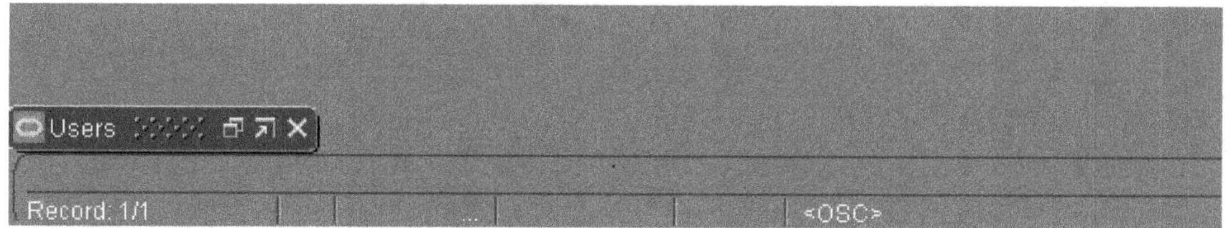

Above screen shot shows that the form has opened in the minimized mode, just above the status bar.

Property: X_POS

This property enables you to setup horizontal position of a form.

We will move the "Users" form little rightward to demonstrate this property.

- Open the Forms Personalization window by clicking on the Help -> Diagnostics -> Custom Code -> Personalize menu option.
    - Seq – 10
    - Description – <Enter meaningful Description>
    - Level – Function
    - Enabled – Yes
- Condition:
    - Trigger Event - WHEN-NEW-FORM-INSTANCE
    - Processing Mode – Not in Enter-Query Mode
- Actions:
    - Seq – 10
    - Type – Property
    - Description – <OPTIONAL>
    - Language – All
    - Enabled – Yes
    - Object Type – Window
    - Target Object- USER_WINDOW

- Property Name: X_POS
- Value – 4 (Initial value was 0)

Save the record.

Property: Y_POS

This property enables you to setup vertical position of a form.

We will move the "Users" form little downward to demonstrate this property.

- Open the Forms Personalization window by clicking on the Help -> Diagnostics -> Custom Code -> Personalize menu option.
    - Seq – 10
    - Description – <Enter meaningful Description>
    - Level – Function
    - Enabled – Yes
- Condition:
    - Trigger Event - WHEN-NEW-FORM-INSTANCE
    - Processing Mode – Not in Enter-Query Mode
- Actions:
    - Seq – 10
    - Type – Property
    - Description – <OPTIONAL>
    - Language – All
    - Enabled – Yes
    - Object Type – Window
    - Target Object- USER_WINDOW
    - Property Name: Y_POS

- Value – 2 (Initial value was 0)

Save the record.

Chapter

5

Object Type: Block

This chapter explains various Personalization properties available for the object type "Block". The properties discussed in this chapter apply to the blocks of a form. Appropriate application of the personalization properties will enable you to establish various data entry validation at the block level while improving the user experience.

Property: ALLOW_NON_SELECTIVE_SEARCH

You might notice, there are few query forms that don't really allow you to perform search unless you mention query criteria value. For example: the "Enter and Maintain" form accessible through People -> Enter and Maintain navigation in the "Human Resources Superuser" or similar responsibility throws as error message when trying to search all people data without specifying any values in the below form.

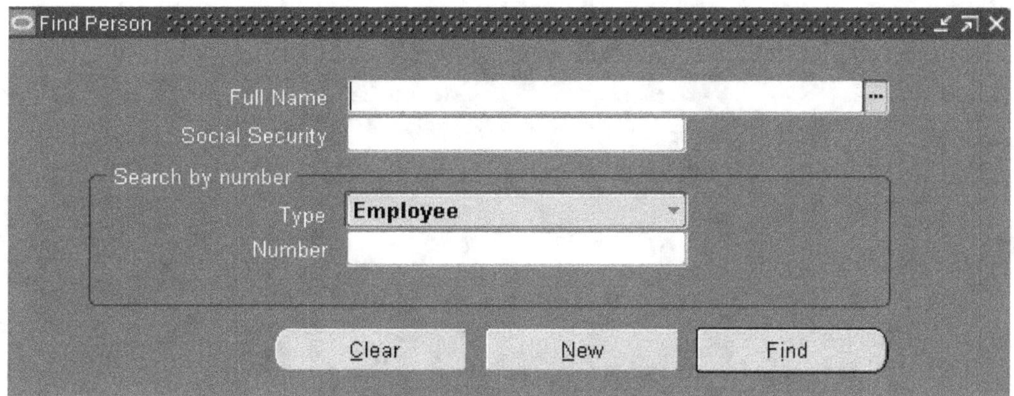

If you click on the "Find" button in the above block, you will get the following error message.

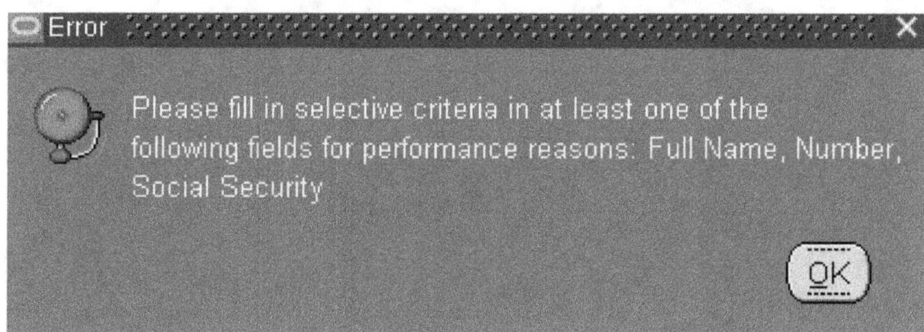

Actually this search restriction has been established to avoid performance issues when searching for all records. Actually the value of the ALLOW_NON_SELECTIVE_SEARCH property has been setup as "FALSE" by default, to enable this restriction.

But this causes bottleneck for many organizations, as they want to see the details of all the employees without specifying any search condition.

We will remove this restriction for the above mentioned block by performing below personalization.

- Open the Forms Personalization window by clicking on the Help -> Diagnostics -> Custom Code -> Personalize menu option.
    - Seq – 10
    - Description – <Enter meaningful Description>

- Level – Function
- Enabled – Yes

- Condition:
  - Trigger Event - WHEN-NEW-FORM-INSTANCE
  - Processing Mode – Not in Enter-Query Mode
- Actions:
  - Seq – 10
  - Type – Property
  - Description – <OPTIONAL>
  - Language – All
  - Enabled – Yes
  - Object Type – Block
  - Target Object- EMP_QF
  - Property Name: ALLOW_NON_SELECTIVE_SEARCH
  - Value – FALSE (Initial value was TRUE)

Save the record.

Property: DEFAULT_WHERE

This property is very useful to establish security in the forms. For example, when we query using F11 and CTRL+F11 in the requisition form (applicable to other forms as well), it actually shows any eligible requisition regardless of the originator. Now, if your business wants to display only the requisitions created by the user who is performing the search, then you may personalize the DEFAULT_WHERE property of the form.

We will demonstrate this property by restricting the requisition display only to a specific requisition number.

- Open the Forms Personalization window by clicking on the Help -> Diagnostics -> Custom Code -> Personalize menu option.

- Seq – 10
- Description – <Enter meaningful Description>
- Level – Function
- Enabled – Yes

- Condition:
  - Trigger Event - WHEN-NEW-FORM-INSTANCE
  - Processing Mode – Not in Enter-Query Mode

- Actions:
  - Seq – 10
  - Type – Property
  - Description – <OPTIONAL>
  - Language – All
  - Enabled – Yes
  - Object Type – Block
  - Target Object- PO_REQ_HDR
  - Property Name: DEFAULT_WHERE
  - Value – segment1 IN ('14368')

Save the record.

This personalization will show only the requisition number – 14368 when you perform query using F11 and CTRL+F11 key combinations.

You may change the value mentioned above to any correct SQL syntax based on the business requirements.

Property: DELETE_ALLOWED

This property allows you to control the ability to DELETE records from a block. Many organizations don't allow deleting records once it is entered into the system.

We will use the purchase requisition form again where you will see that it allows users to delete a requisition, as the delete icon is enabled in the above toolbar.

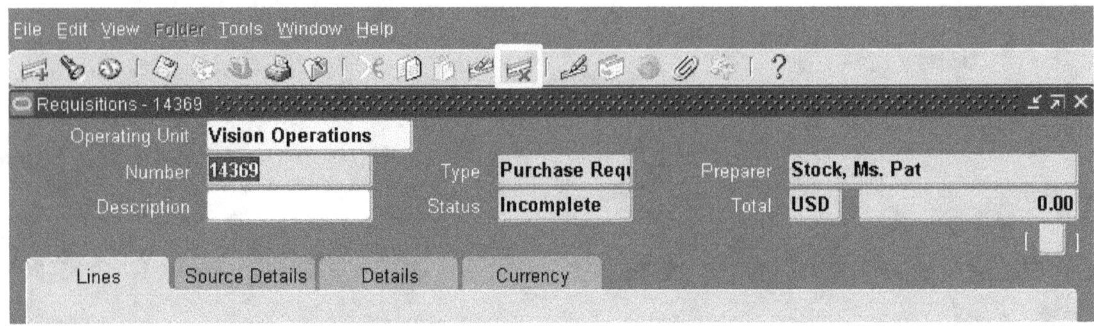

We will perform a forms personalization using the DELETE_ALLOWED property to restriction deletion in the requisition form.

- Open the Forms Personalization window by clicking on the Help -> Diagnostics -> Custom Code -> Personalize menu option.
    - Seq – 10
    - Description – <Enter meaningful Description>
    - Level – Function
    - Enabled – Yes
- Condition:
    - Trigger Event - WHEN-NEW-FORM-INSTANCE
    - Processing Mode – Not in Enter-Query Mode
- Actions:
    - Seq – 10
    - Type – Property
    - Description – <OPTIONAL>
    - Language – All
    - Enabled – Yes
    - Object Type – Block

- Target Object- PO_REQ_HDR
- Property Name: DELETE_ALLOWED
- Value – FALSE

Save the record.

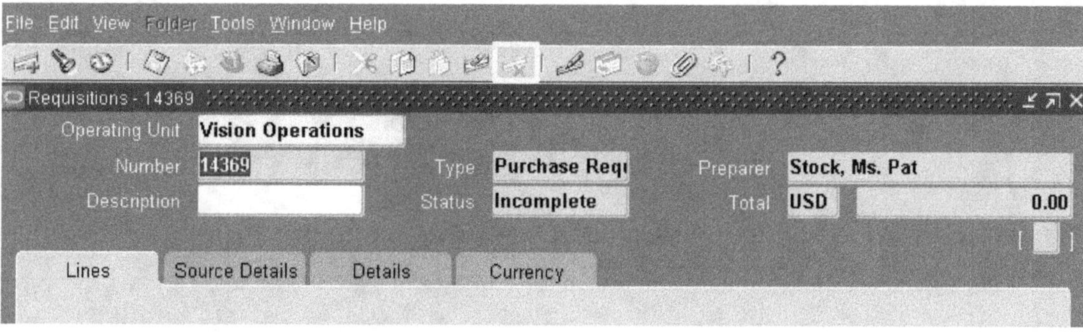

Now the Delete icon is disabled.

Property: EXPORT_HIDDEN_CANVASES

This is an amazing personalization property that has been rolled out Release 12 onwards which allows users to export data to Spreadsheet even from the hidden canvases of a form. For example, if you click on the File -> Export menu to export requisition data data to Spreadsheet in the below requisition form, it will export data from the "Lines" tab only.

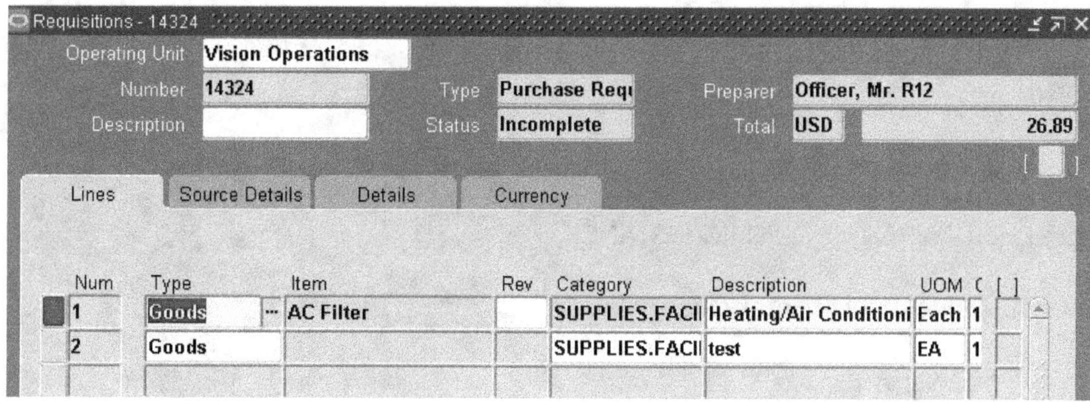

The exported data is shown below.

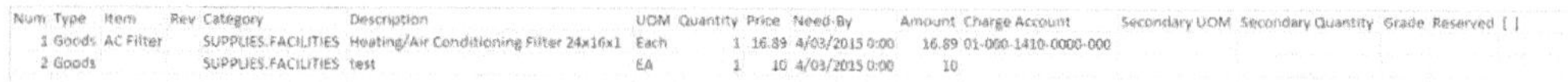

Num	Type	Item	Rev	Category	Description	UOM	Quantity	Price	Need-By	Amount	Charge Account	Secondary UOM	Secondary Quantity	Grade	Reserved	[ ]
1	Goods	AC Filter		SUPPLIES.FACILITIES	Heating/Air Conditioning Filter 24x16x1	Each	1	16.89	4/03/2015 0:00	16.89	01-000-1410-0000-000					
2	Goods			SUPPLIES.FACILITIES	test	EA	1	10	4/03/2015 0:00	10						

There are data in the other tabs (e.g. Currency) too but it doesn't really get exported to Spreadsheet. I have seen many users complaining about it in the previous versions, as they can't really fully report or analyze data using the Export functionality.

The “Currency” tab/canvas contains below data which couldn’t be exported.

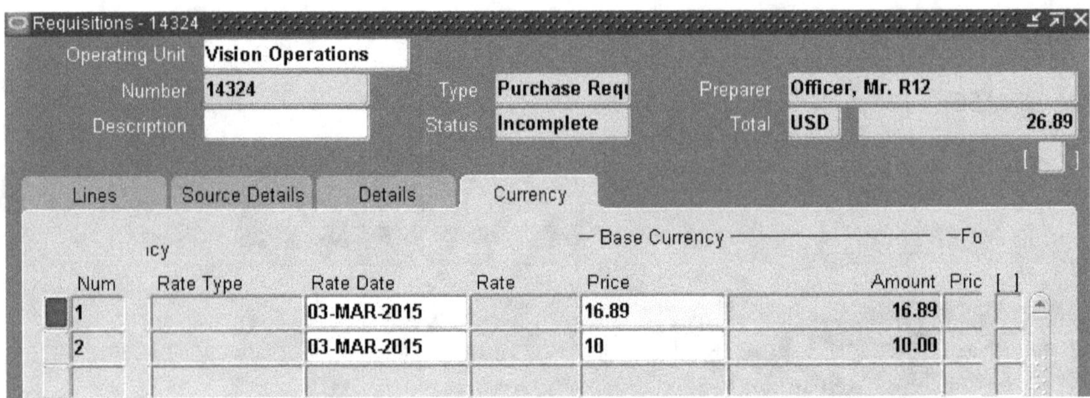

Oracle has provided a very effective resolution to the above reported user frustration by rolling out the EXPORT_HIDDEN_CANVASES personalization property.

We will demonstrate this property by enabling data export from all the tabs/canvases of the requisition lines block displayed above.

- Open the Forms Personalization window by clicking on the Help -> Diagnostics -> Custom Code -> Personalize menu option.
    - Seq – 10
    - Description – <Enter meaningful Description>
    - Level – Function
    - Enabled – Yes
- Condition:
    - Trigger Event - WHEN-NEW-FORM-INSTANCE
    - Processing Mode – Not in Enter-Query Mode
- Actions:
    - Seq – 10
    - Type – Property
    - Description – <OPTIONAL>
    - Language – All

- Enabled – Yes
- Object Type – Block
- Target Object- LINES
- Property Name: EXPORT_HIDDEN_CANVASES
- Value – TRUE

Save the record.

Now the exported data looks like below.

Num	Type	Item	Rev	Category	Description	UOM	Quantity	Price	Need-By	Amount	Charge Account	Rate Type	Rate Date	Rate Date	Rate	Base Currency: Price	Base Currency: Amount	Foreign Currency:
1	Goods	AC Filter		SUPPLIES.FACILITIES	Heating/Air Conditioning Filter 24x16x1	Each	1	16.89	4/03/2015 0:00	16.89	01-000-1410-0000-000		3-Mar-15			16.89	16.89	
2	Goods			SUPPLIES.FACILITIES	test	EA	1	10	4/03/2015 0:00	10			3-Mar-15			10	10	

You might notice that, the data from the “Currency” canvas also appears now in the Spreadsheet. Actually there are many columns that appear between the “Charge Account” and “Rate Type” columns but I had to hide them to show you the data exported from the Currency canvas.

Property: INSERT_ALLOWED

This property allows you to control the ability of inserting records into a form block. There are business scenarios where the management allows certain group of users to insert the header level records but not the line level details. We will restrict insertion into the requisition lines block to demonstrate this property.

- Open the Forms Personalization window by clicking on the Help -> Diagnostics -> Custom Code -> Personalize menu option.
    - Seq – 10
    - Description – <Enter meaningful Description>
    - Level – Function
    - Enabled – Yes
- Condition:
    - Trigger Event - WHEN-NEW-FORM-INSTANCE
    - Processing Mode – Not in Enter-Query Mode
- Actions:

- Seq – 10
- Type – Property
- Description – <OPTIONAL>
- Language – All
- Enabled – Yes
- Object Type – Block
- Target Object- LINES
- Property Name: INSERT_ALLOWED
- Value – FALSE

Save the record.

Property: ORDER_BY

This is very frequently used property, as this allows users to sort a block level data based on their preference.

For example, the requisition line level data appears based on line number sequence by default (as shown below). The line quantities are 200,200 and 11 respectively. Now, the user wants this data to be sorted based on line quantity. We can achieve this requirement by performing following personalization.

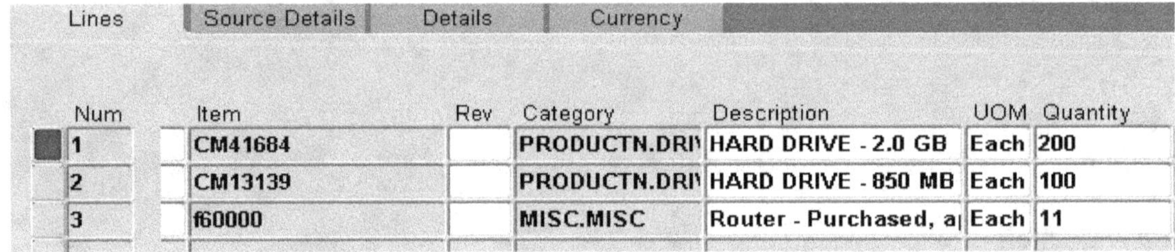

Lines | Source Details | Details | Currency

Num	Item	Rev	Category	Description	UOM	Quantity
1	CM41684		PRODUCTN.DRI\	HARD DRIVE - 2.0 GB	Each	200
2	CM13139		PRODUCTN.DRI\	HARD DRIVE - 850 MB	Each	100
3	f60000		MISC.MISC	Router - Purchased, a|	Each	11

- Open the Forms Personalization window by clicking on the Help -> Diagnostics -> Custom Code -> Personalize menu option.
  - Seq – 10
  - Description – <Enter meaningful Description>
  - Level – Form
  - Enabled – Yes

- Condition:
    - Trigger Event - WHEN-NEW-BLOCK-INSTANCE
    - Processing Mode – Not in Enter-Query Mode
- Actions:
    - Seq – 10
    - Type – Property
    - Description – <OPTIONAL>
    - Language – All
    - Enabled – Yes
    - Object Type – Block
    - Target Object- LINES
    - Property Name: ORDER_BY
    - Value – QUANTITY

Save the record.

Now the records appear as below.

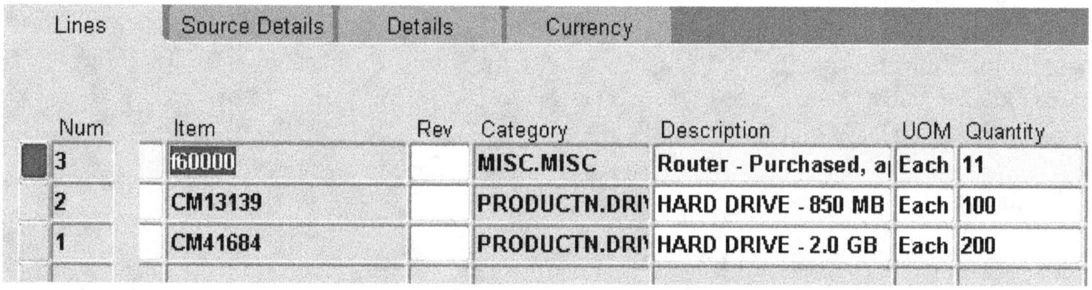

Lines | Source Details | Details | Currency

Num	Item	Rev	Category	Description	UOM	Quantity
3	f60000		MISC.MISC	Router - Purchased, a	Each	11
2	CM13139		PRODUCTN.DRI	HARD DRIVE - 850 MB	Each	100
1	CM41684		PRODUCTN.DRI	HARD DRIVE - 2.0 GB	Each	200

Property: QUERY_ALLOWED

As the name suggests, this property allows you to control the ability to query records in a data block. Usually users are able to query all the records in a form regardless of the originator of the record. It is an issue for the organization or business unit that deals with sensitive data; therefore such organization might restrict the ability to query data created by other users.

We will restrict the ability to query records in the requisition lines block to demonstrate this property.

- Open the Forms Personalization window by clicking on the Help -> Diagnostics -> Custom Code -> Personalize menu option.
    - Seq – 10
    - Description – <Enter meaningful Description>
    - Level – Form
    - Enabled – Yes
- Condition:
    - Trigger Event - WHEN-NEW-FORM-INSTANCE
    - Processing Mode – Not in Enter-Query Mode
- Actions:
    - Seq – 10
    - Type – Property
    - Description – <OPTIONAL>
    - Language – All
    - Enabled – Yes
    - Object Type – Block
    - Target Object- LINES
    - Property Name: QUERY_ALLOWED
    - Value – FALSE

Save the record.

Now the users won't be able to query requisition line level records.

Property: UPDATE_ALLOWED

This property allows you to control the ability to update records in a particular block. For example, the records in the requisition lines block are updatable by default when you query an "Incomplete" requisition as shown below.

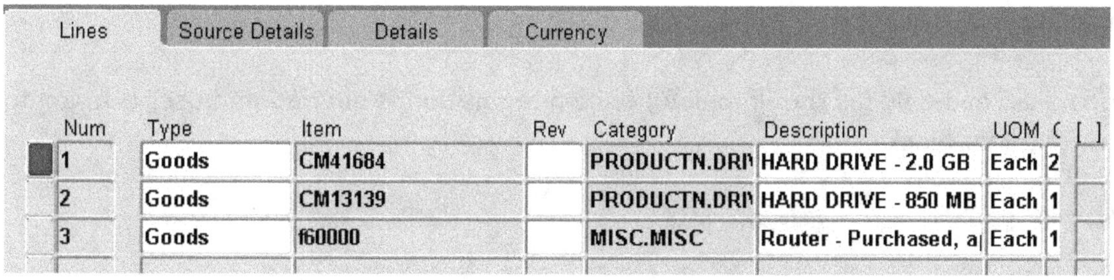

Lines | Source Details | Details | Currency

Num	Type	Item	Rev	Category	Description	UOM	(	[ ]
1	Goods	CM41684		PRODUCTN.DRI	HARD DRIVE - 2.0 GB	Each	2	
2	Goods	CM13139		PRODUCTN.DRI	HARD DRIVE - 850 MB	Each	1	
3	Goods	f60000		MISC.MISC	Router - Purchased, a	Each	1	

Following personalization steps will make the above block read-only.

- Open the Forms Personalization window by clicking on the Help -> Diagnostics -> Custom Code -> Personalize menu option.
    - Seq – 10
    - Description – <Enter meaningful Description>
    - Level – Form
    - Enabled – Yes
- Condition:
    - Trigger Event - WHEN-NEW-BLOCK-INSTANCE
    - Trigger Object - LINES
    - Processing Mode – Not in Enter-Query Mode
- Actions:
    - Seq – 10
    - Type – Property
    - Description – <OPTIONAL>
    - Language – All
    - Enabled – Yes
    - Object Type – Block
    - Target Object- LINES
    - Property Name: UPDATE_ALLOWED

- Value – FALSE

Now the block appears as read-only (as shown below) once a requisition is queried and user is trying to update the records in the lines block.

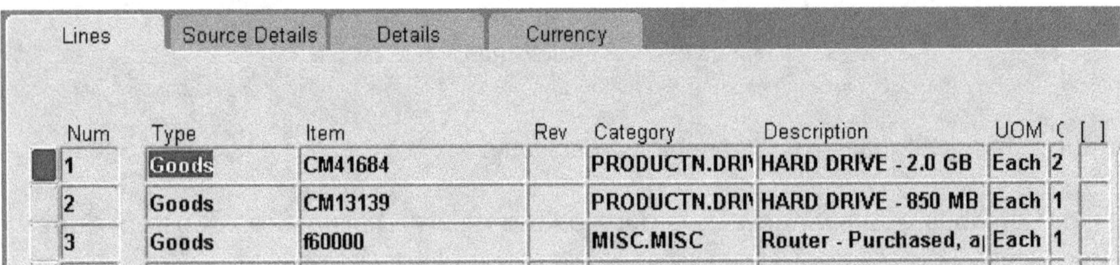

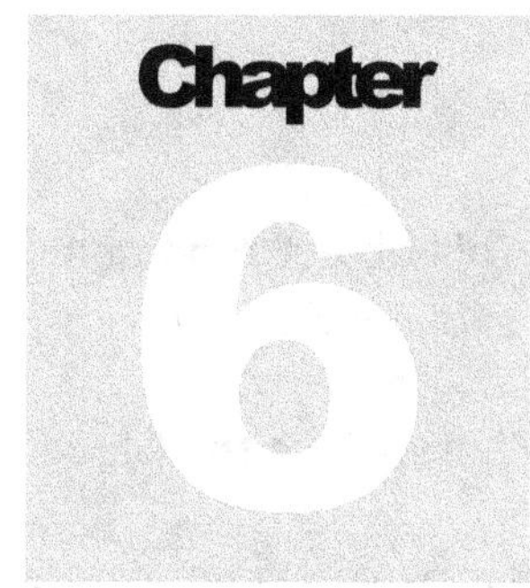

Object Type: Tab Page

This chapter explains various Personalization properties available for the "Tab Page" object type. The properties discussed in this chapter will help you to tailor the tab pages as per the business requirements.

This object type deals with the tab pages of a form and allows you to control multiple characteristics as explained below.

We will use the Move Order form to explain this object type.

Once you open the Move Order form, you will have the ability to query move order headers and lines based on a set of search criteria in addition to the CTRL and CTRL+F11 method. This is done by clicking on the "Find" icon available in the toolbar (as shown below).

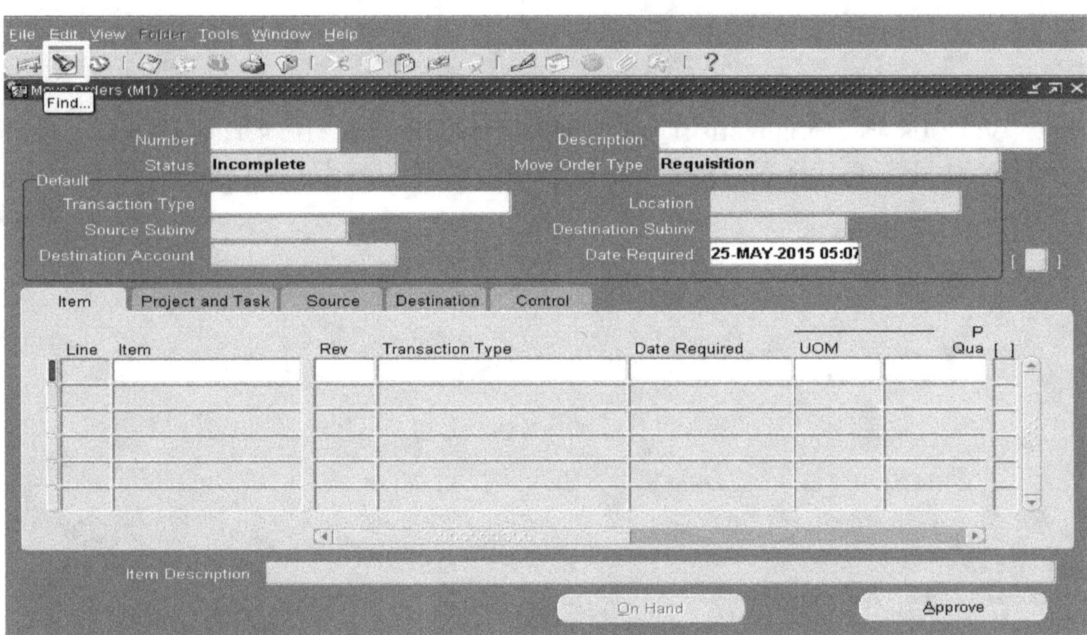

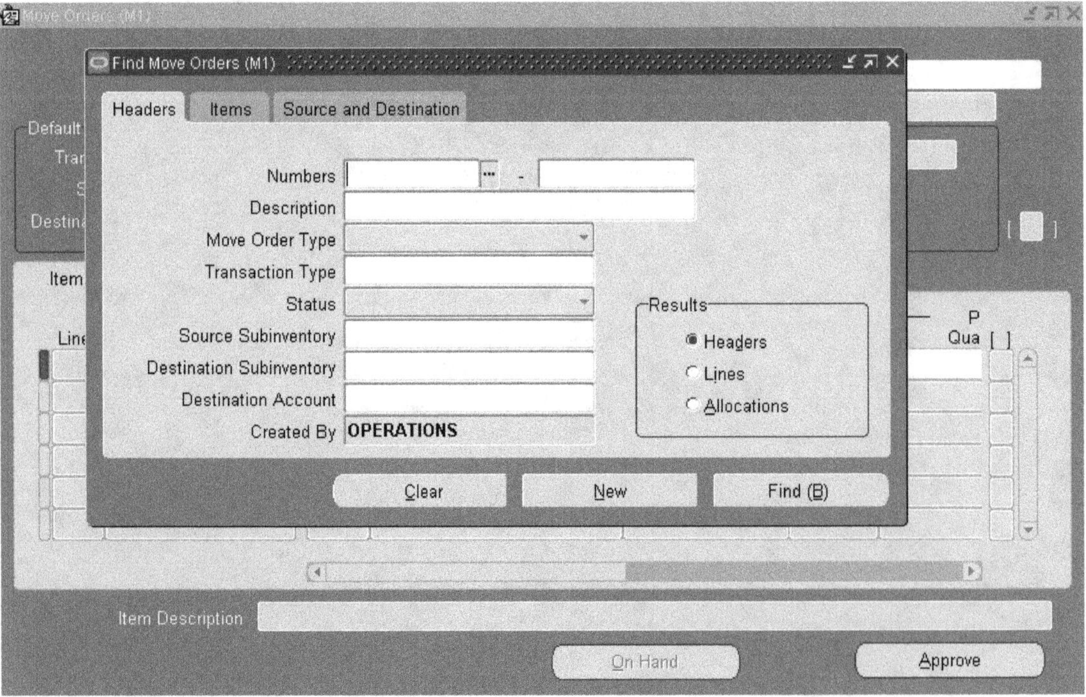

Once you click on the "Find" icon, above search window appears which has 3 tab pages as "Headers", "Items" and "Source and Destination". We will change the characteristics of these tab pages to explain the properties available under "Tab Page" object type.

Property: DISPLAYED

As the name suggests, this property allows you to control the visibility of a tab page. Following personalization steps will remove the "Headers" tab page from the above search window.

- Open the Forms Personalization window by clicking on the Help -> Diagnostics -> Custom Code -> Personalize menu option.
    - Seq – 10
    - Description – <Enter meaningful Description>
    - Level – Form
    - Enabled – Yes
- Condition:
    - Trigger Event - WHEN-NEW-FORM-INSTANCE
    - Processing Mode – Not in Enter-Query Mode
- Actions:
    - Seq – 10
    - Type – Property
    - Description – <OPTIONAL>
    - Language – All
    - Enabled – Yes
    - Object Type – Tab Page
    - Target Object- TOMAI_QF_MAIN_LIST.TOMAI_QF_HEADER_CAN
    - Property Name: DISPLAYED
    - Value – FALSE

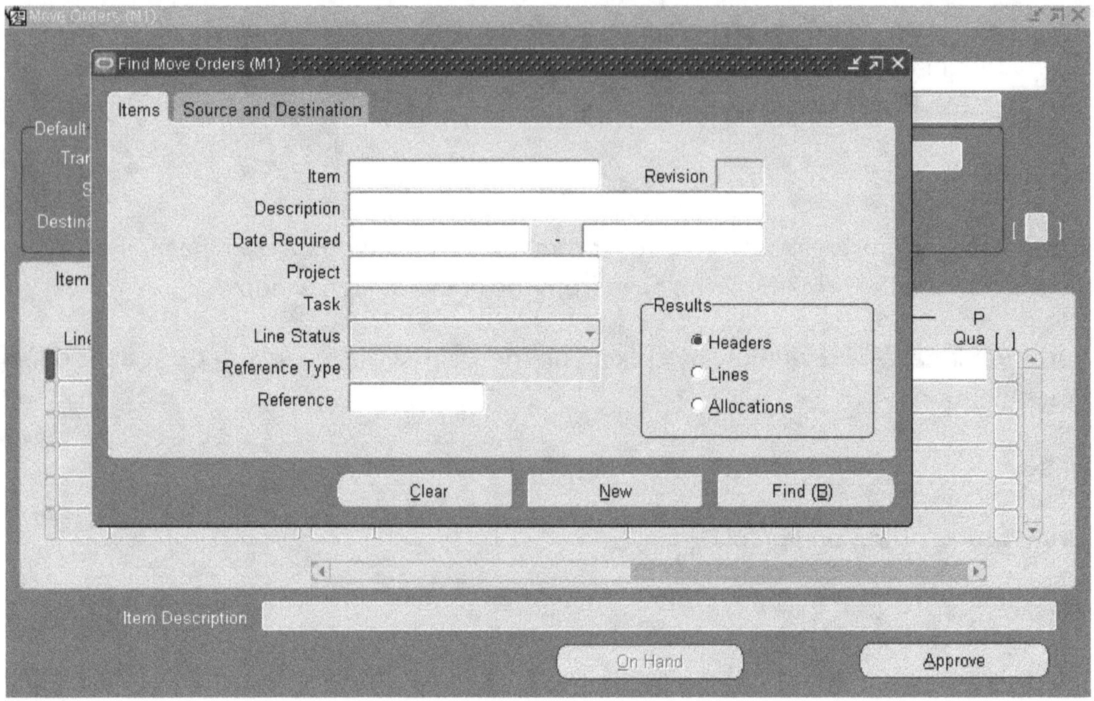

The “Headers” tab disappeared from the above window.

Property: ENABLED

You may use this personalization property when you want to disable a tab page when it’s enabled and vice-versa.

We will disable the “Items” tab page to demonstrate this personalization property.

- Open the Forms Personalization window by clicking on the Help -> Diagnostics -> Custom Code -> Personalize menu option.
    - Seq – 10
    - Description – <Enter meaningful Description>
    - Level – Form
    - Enabled – Yes
- Condition:
    - Trigger Event - WHEN-NEW-FORM-INSTANCE
    - Processing Mode – Not in Enter-Query Mode

- Actions:
    - Seq – 10
    - Type – Property
    - Description – <OPTIONAL>
    - Language – All
    - Enabled – Yes
    - Object Type – Tab Page
    - Target Object- TOMAI_QF_MAIN_LIST.TOMAI_QF_ITEM_CAN
    - Property Name: ENABLED
    - Value – FALSE

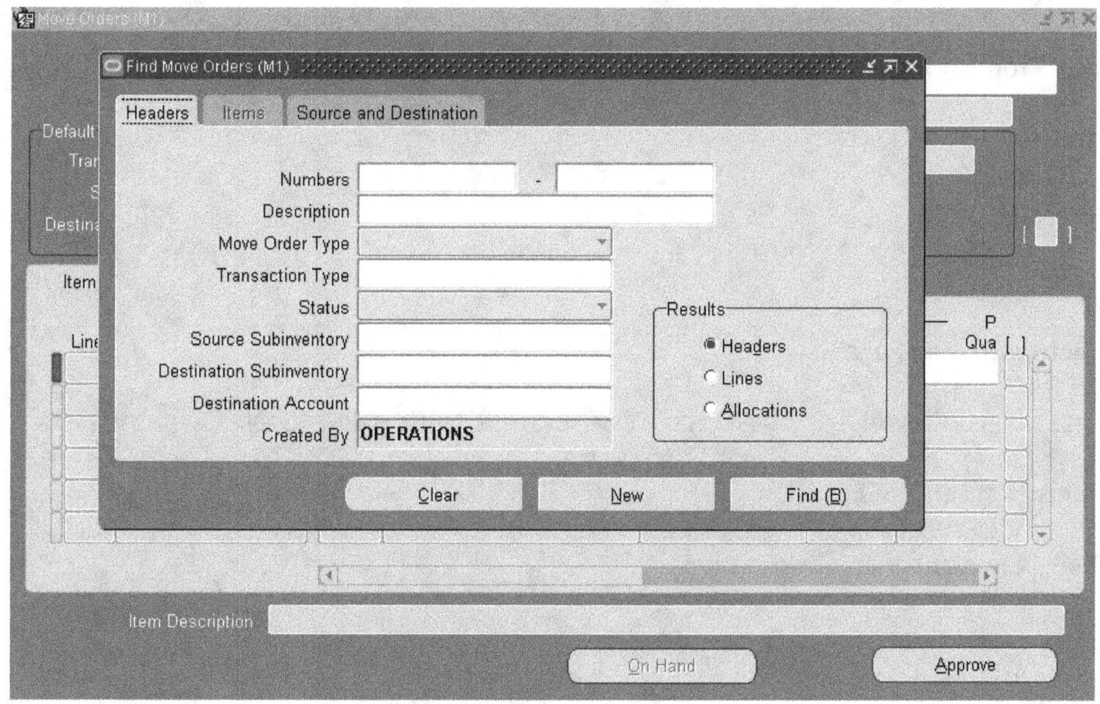

Above screen shot shows the disabled "Items" tab page.

Property: LABEL

This property is used to change the caption/label of a tab page. We will change the label of the "Source and Destination" tab page to "Item Source and Destination" to demonstrate this personalization.

- Open the Forms Personalization window by clicking on the Help -> Diagnostics -> Custom Code -> Personalize menu option.
  - Seq – 10
  - Description – <Enter meaningful Description>
  - Level – Form
  - Enabled – Yes
- Condition:
  - Trigger Event - WHEN-NEW-FORM-INSTANCE
  - Processing Mode – Not in Enter-Query Mode
- Actions:
  - Seq – 10
  - Type – Property
  - Description – <OPTIONAL>
  - Language – All
  - Enabled – Yes
  - Object Type – Tab Page
  - Target Object- TOMAI_QF_MAIN_LIST.TOMAI_QF_SOURCE_DEST_CAN
  - Property Name: LABEL
  - Value – Item Source and Destination

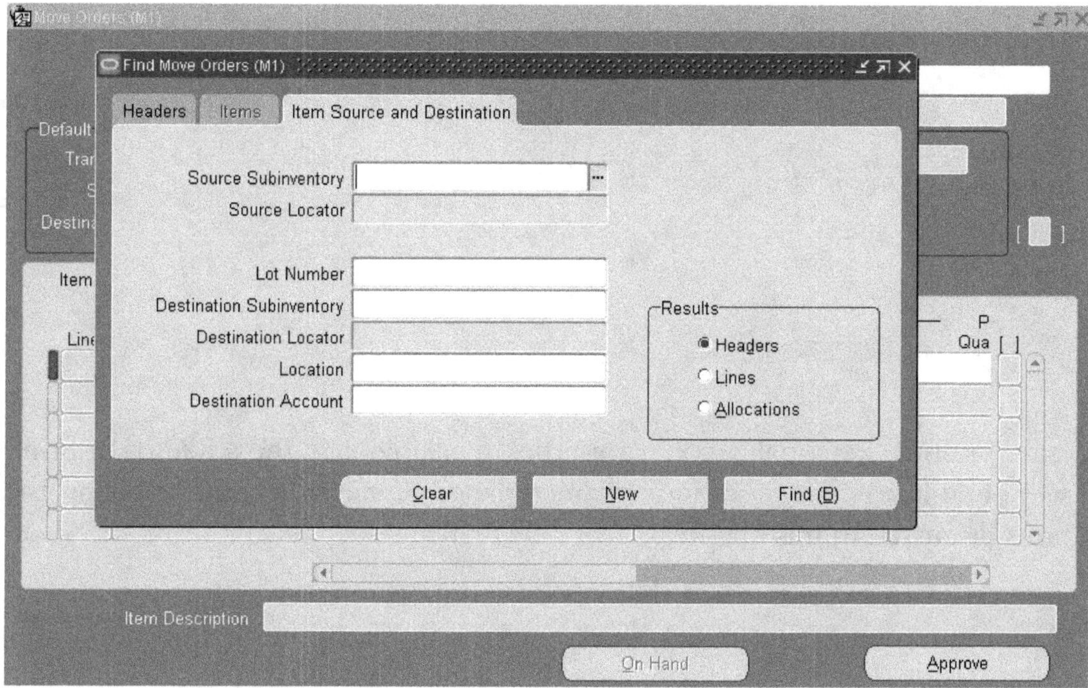

Above screen shot shows the modified label of the tab page.

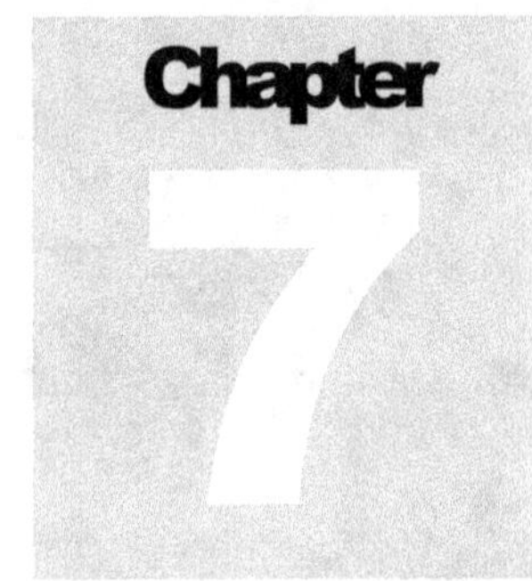

Object Type: Radio Button

This chapter explains various Personalization properties available for the "Radio Button" object type. You will be able to change various characteristics of radio buttons applying the forms personalizations discussed in this chapter.

This object type deals with the radio buttons. Not every form will have the radio buttons but wherever it's available, you can control the characteristics of such buttons using the properties available under this object type.

We will use the radio buttons available in the Move Order search window to demonstrate the properties.

Property: ENABLED

This property allows you to enable/disable a radio button. We will disable the "Headers" radio button using following personalization steps.

- Open the Forms Personalization window by clicking on the Help -> Diagnostics -> Custom Code -> Personalize menu option.
    - Seq – 10
    - Description – <Enter meaningful Description>
    - Level – Form
    - Enabled – Yes
- Condition:
    - Trigger Event - WHEN-NEW-FORM-INSTANCE
    - Processing Mode – Not in Enter-Query Mode
- Actions:
    - Seq – 10
    - Type – Property
    - Description – <OPTIONAL>
    - Language – All
    - Enabled – Yes
    - Object Type – Radio Button
    - Target Object- TOMAI_QF_BLK.VIEW_RESULTS.HEADER
    - Property Name: ENABLED

- Value – False

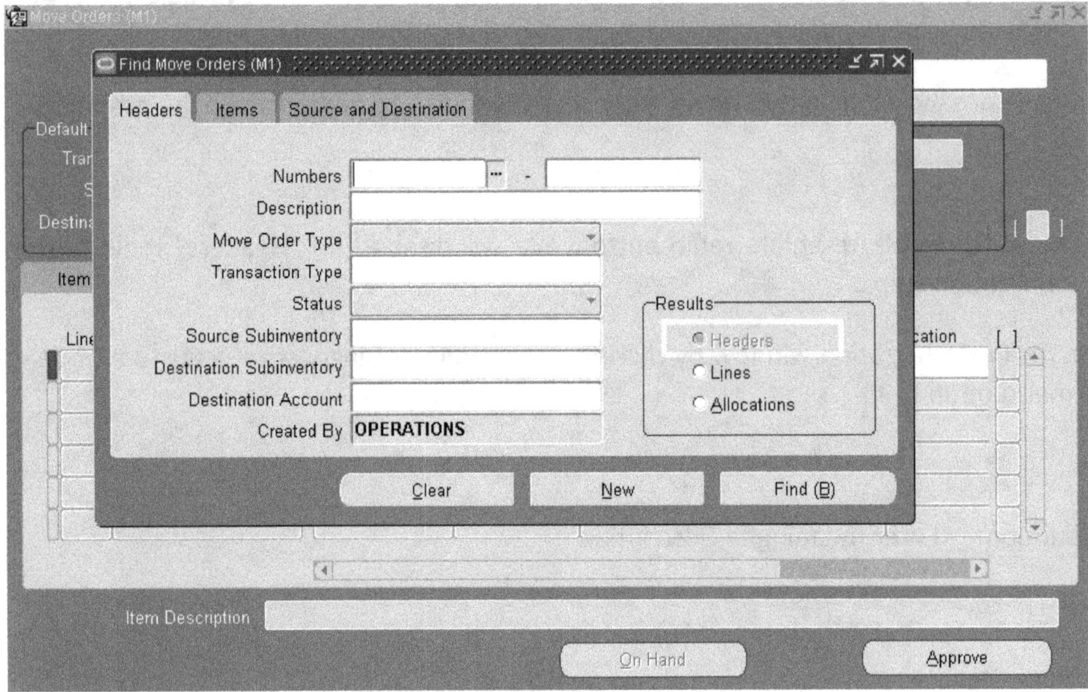

Above screen shot shows the disabled radio button.

Property: LABEL

This property can be used to change the label of a radio button. We will change the label of the "Allocations" radio button to "Reservation" to demonstrate the property.

- Open the Forms Personalization window by clicking on the Help -> Diagnostics -> Custom Code -> Personalize menu option.
  - Seq – 10
  - Description – <Enter meaningful Description>
  - Level – Form
  - Enabled – Yes
- Condition:
  - Trigger Event - WHEN-NEW-FORM-INSTANCE
  - Processing Mode – Not in Enter-Query Mode

- Actions:
    - Seq – 10
    - Type – Property
    - Description – <OPTIONAL>
    - Language – All
    - Enabled – Yes
    - Object Type – Radio Button
    - Target Object- TOMAI_QF_BLK.VIEW_RESULTS.LINE_DTL
    - Property Name: LABEL
    - Value – Reservat&ions (initial value Allocat&ions)

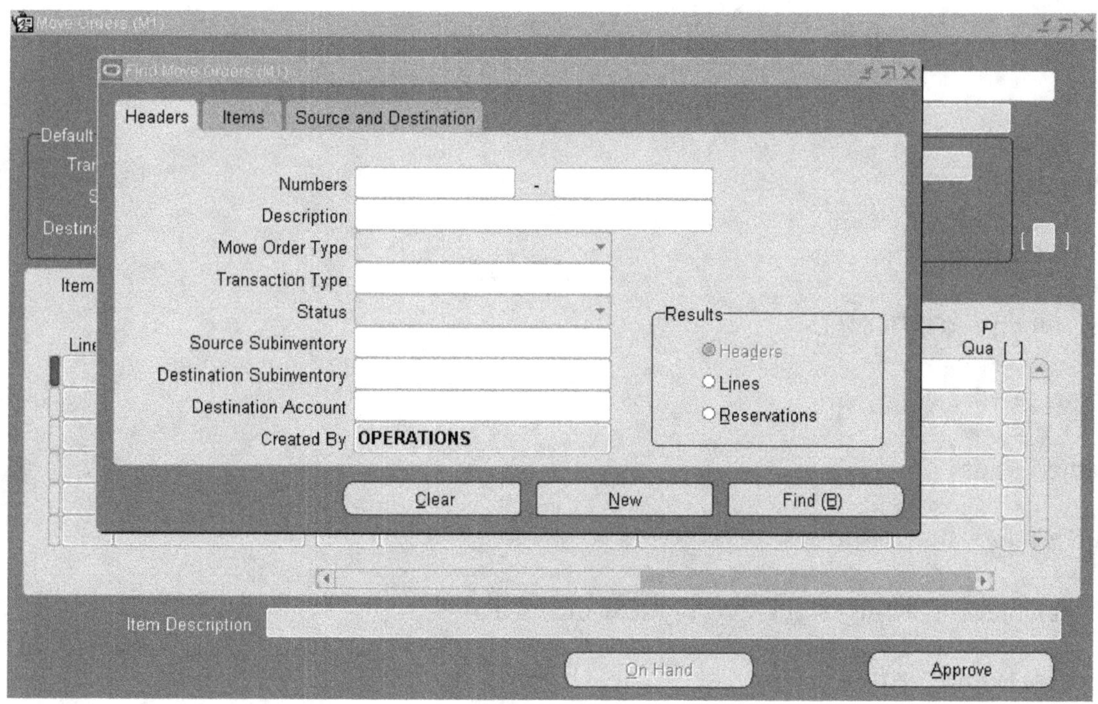

Above screen shot shows the changed label.

Property: VISIBLE

You can show/hide a radio button using this personalization property. We will hide the "Lines" radio button appears in the previous screen shot to demonstrate this property.

- Open the Forms Personalization window by clicking on the Help -> Diagnostics -> Custom Code -> Personalize menu option.
    - Seq – 10
    - Description – <Enter meaningful Description>
    - Level – Form
    - Enabled – Yes
- Condition:
    - Trigger Event - WHEN-NEW-FORM-INSTANCE
    - Processing Mode – Not in Enter-Query Mode
- Actions:
    - Seq – 10
    - Type – Property
    - Description – <OPTIONAL>
    - Language – All
    - Enabled – Yes
    - Object Type – Radio Button
    - Target Object- TOMAI_QF_BLK.VIEW_RESULTS.LINES
    - Property Name: VISIBLE
    - Value – FALSE

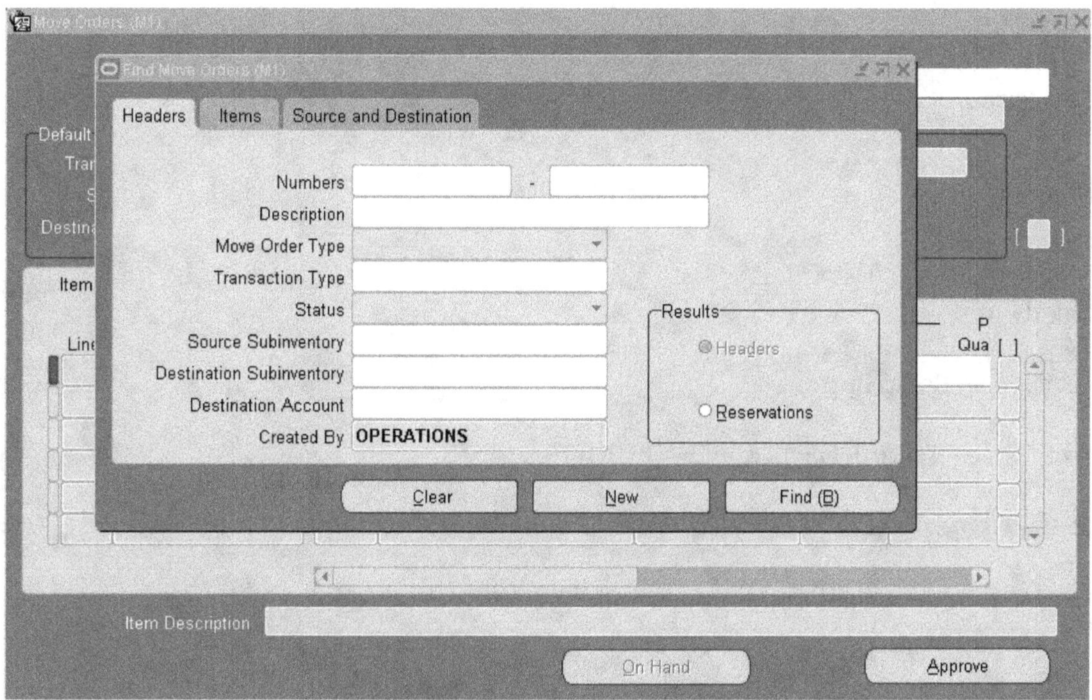

The "Lines" radio button has disappeared from the above screen shot.

Property: X_POS

This property allows you to setup the horizontal position of a radio button in a form/canvas. We will move the "Lines" radio button towards right to demonstrate this property.

- Open the Forms Personalization window by clicking on the Help -> Diagnostics -> Custom Code -> Personalize menu option.
    - Seq – 10
    - Description – <Enter meaningful Description>
    - Level – Form
    - Enabled – Yes
- Condition:
    - Trigger Event - WHEN-NEW-FORM-INSTANCE
    - Processing Mode – Not in Enter-Query Mode
- Actions:

- Seq – 10
- Type – Property
- Description – <OPTIONAL>
- Language – All
- Enabled – Yes
- Object Type – Radio Button
- Target Object- TOMAI_QF_BLK.VIEW_RESULTS.LINES
- Property Name: X_POS
- Value – 0.8 (initial value 0.5)

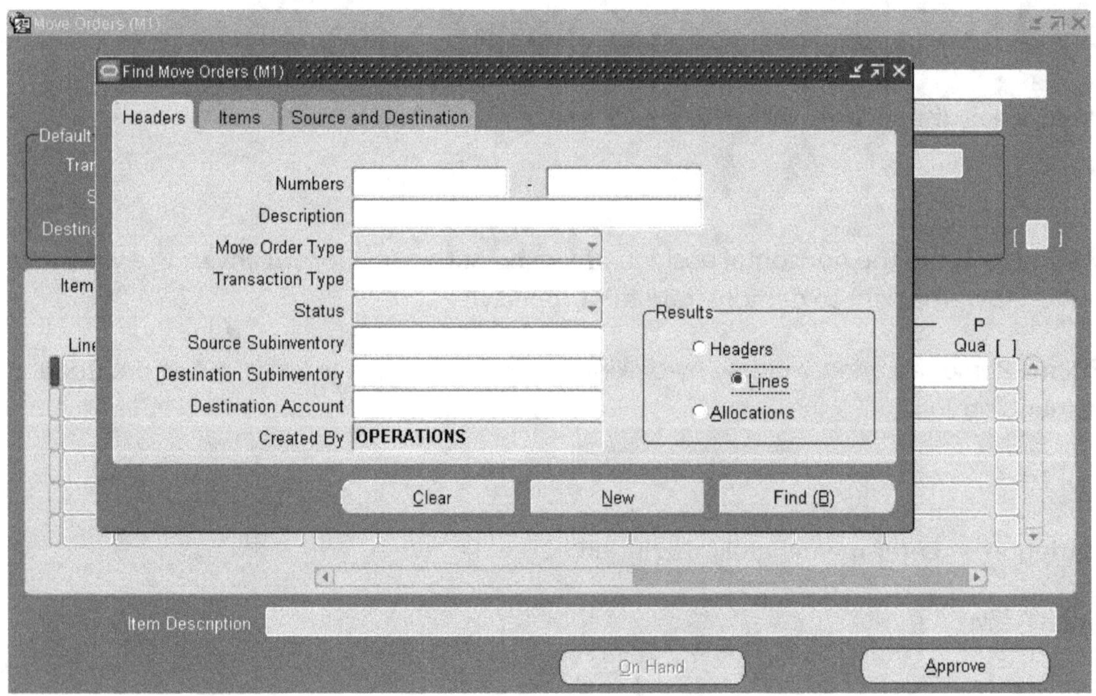

Above screen shot shows the change.

Property: Y_POS

This property allows you to setup the vertical position of a radio button in a form/canvas. We will move the "Allocations" radio button little downward to demonstrate this property.

- Open the Forms Personalization window by clicking on the Help -> Diagnostics -> Custom Code -> Personalize menu option.
    - Seq – 10
    - Description – <Enter meaningful Description>
    - Level – Form
    - Enabled – Yes
- Condition:
    - Trigger Event - WHEN-NEW-FORM-INSTANCE
    - Processing Mode – Not in Enter-Query Mode
- Actions:
    - Seq – 10
    - Type – Property
    - Description – <OPTIONAL>
    - Language – All
    - Enabled – Yes
    - Object Type – Radio Button
    - Target Object- TOMAI_QF_BLK.VIEW_RESULTS.LINE_DTL
    - Property Name: Y_POS
    - Value – 0.86 (initial value 0.76)

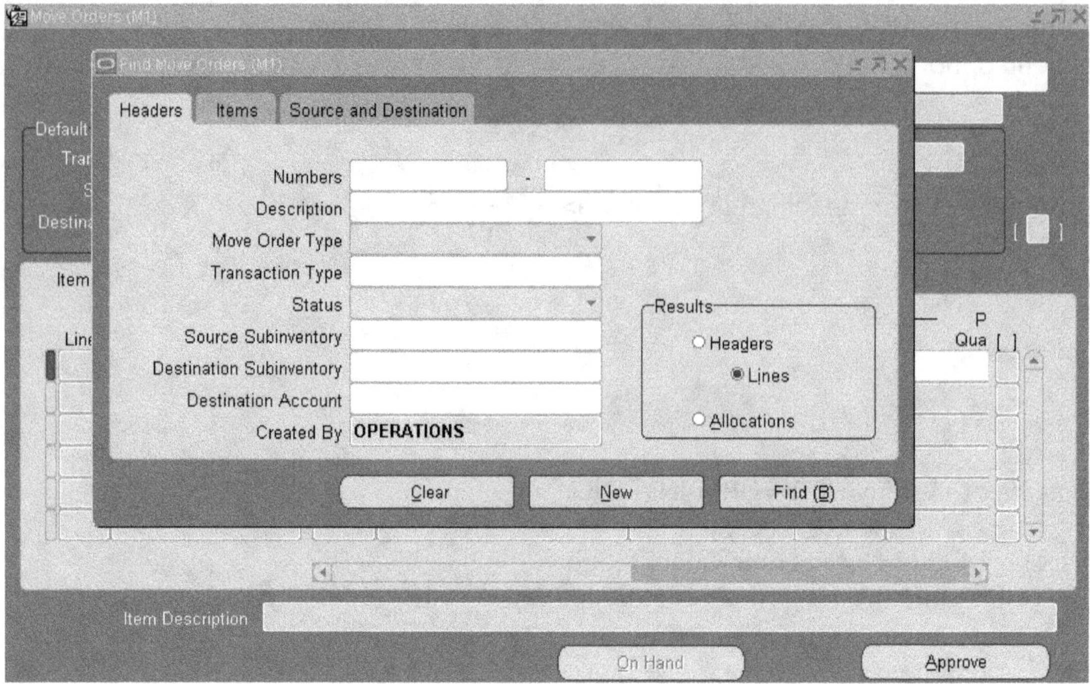

Above screen shot shows the change.

Object Type: Global Variable

This chapter explains various Personalization properties available for the object type "Global Variable". This is one of the more powerful object types of Forms Personalization and you will be able to utilize the power of this object type once you complete reading this chapter.

The capability of the Oracle Forms Personalization can be extended further through the Global Variables. You can deliver numerous functionalities for the business using Global Variables. Following real life business scenario will be used to demonstrate this amazing personalization object type.

Please be mindful that, use Global variable only when you need to refer the value of the variable to multiple forms, otherwise use local variable (explained late in this book) if your business logic is limited to a single form.

Business Scenario: We will extend the example mentioned in the "MENUn and SPECIALn" section of the "Triggers" chapter earlier and explore how to auto populate the purchase order number in the "Purchase Order Numbers From" and "To" field of the "PO Output for Communication" program using Global Variables.

Just to reiterate, we wanted user to launch the "PO Output for Communication" program directly from the purchase order form by clicking a menu item and auto populate the purchase order number in the parameters mentioned above, so that user doesn't need to follow the View -> Request procedure to submit the program.

Though most of the below setups were explained in the "MENUn and SPECIALn" section of the "Triggers" chapter earlier, we are repeating the same to explain the Global Variable component. Highlighted section has been added to incorporate Global Variables.

Seq – 10

Description – Define "Launch PDF" submenu.

Level – Function

Enabled – Yes

Condition:

Trigger Event: WHEN-NEW-FORM-INSTANCE (Note: Sub menus can be defined only in the WHEN-NEW-FORM-INSTANCE trigger.)

Condition: <You may enter your condition here>

Processing Mode: Not in Enter-Query Mode.

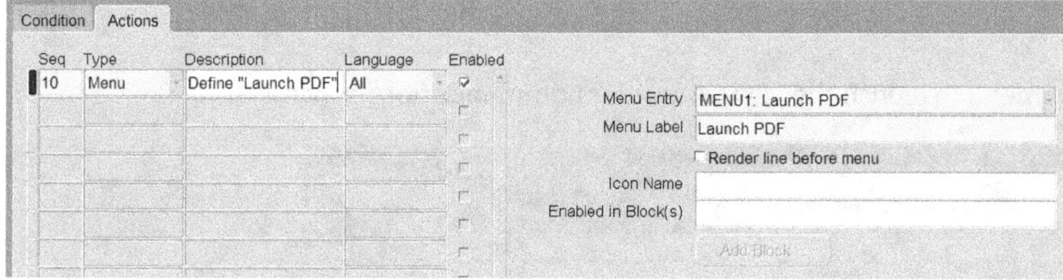

Actions:

Seq: 10

Type: Menu

Description: Define "Launch PDF"

Language: All

Enabled: Yes

Menu Entry: MENU1

Menu Label: Launch PDF

Now you have completed the setups required to create the submenu "Launch PDF" under the Tools menu as per below screen shot.

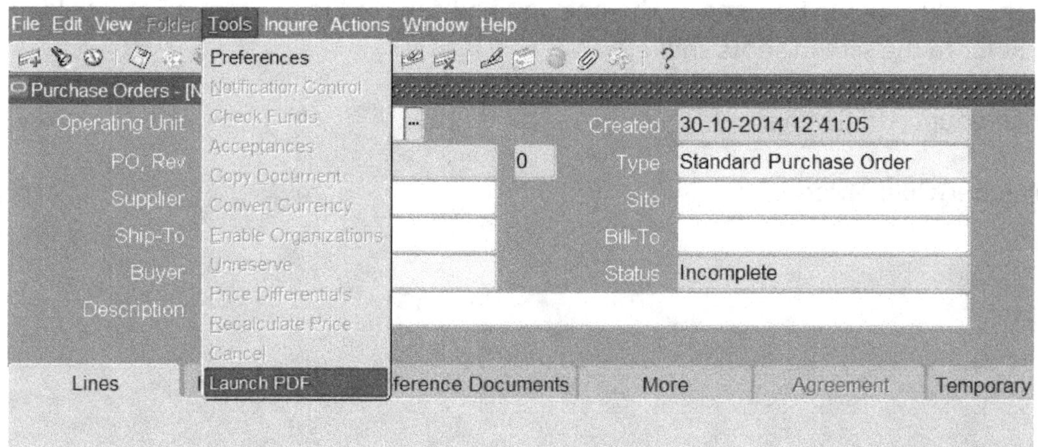

At this stage, nothing really will happen if you click on this submenu. The objective here is to launch the "PO Output for Communication" concurrent program to view the PDF. We need to perform another personalization to achieve this.

Step-2:

Enable “Launch PDF” submenu to launch the “PO Output for Communication” program.

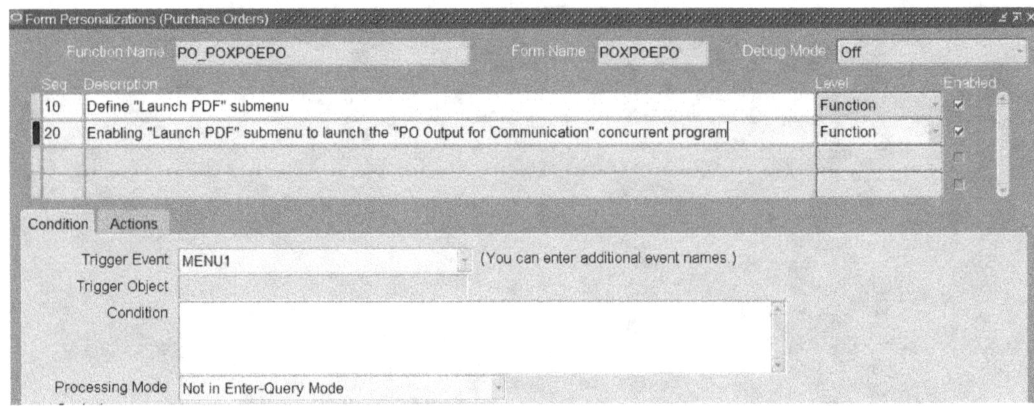

Now adding the sequence 20.

Seq: 20

Description: Enabling "Launch PDF" submenu to launch the "PO Output for Communication" concurrent program

Level: Function

Enabled: Yes

Condition:

Trigger Event: MENU1 (This value should be same as the one you had selected in the Menu Entry field of Seq-10 personalization. Refer to the screen shots above.)

Condition: <You may enter your condition here>

Processing Mode: Not in Enter-Query Mode.

Actions:

Seq: 5

Type: Property

Description: Assigning the Global Variable

Language: All

Enabled: Yes

Object Type: Global Variable

Target Object: G_PO_NUM

Property Name: VALUE

VALUE: =:PO_HEADERS.SEGMENT1 (This syntax assigns the PO number directly from the form to the variable).

Seq: 10

Type: Builtin

Description: Launching the Program

Language: All

Enabled: Yes

Builtin Type: Launch SRS Form

Program Name: PO Output for Communication

Save the setup.

Once the above setup is done, you need to perform minor modifications to the setup of the "PO Output for Communication" concurrent program as explained below.

Responsibility: Application Developer or similar

Navigation: Concurrent -> Program

Query the concurrent program "PO Output for Communication".

Click on the ""Parameters" button.

Locate the "P_po_num_from" parameter as shown below.

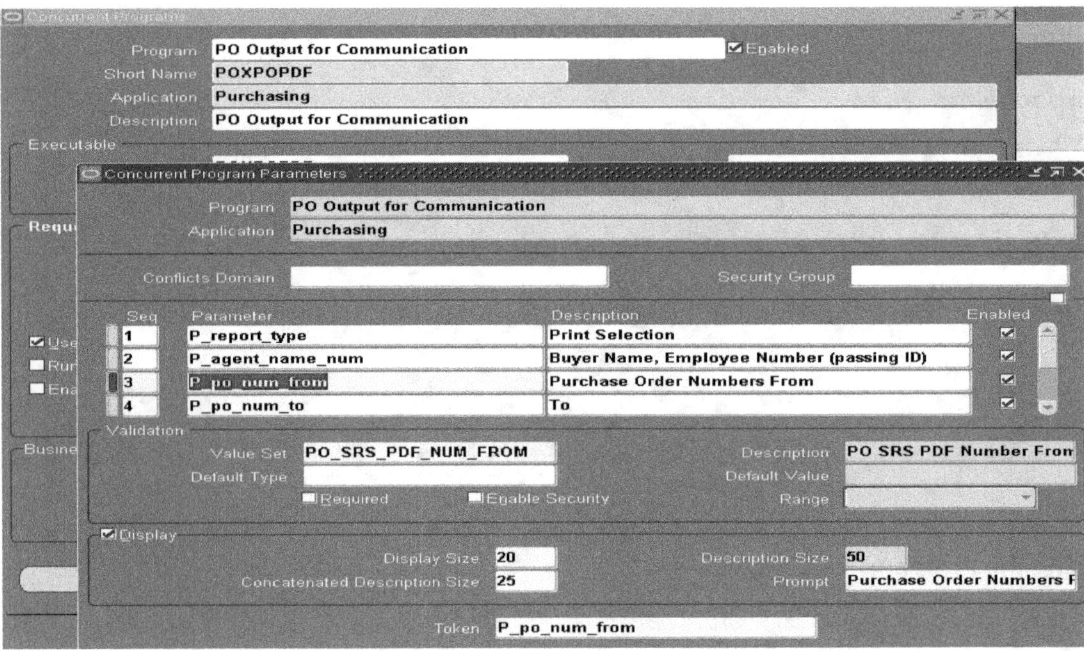

Update the Default Type to "SQL Statement" and Default Value to "SELECT :GLOBAL.G_PO_NUM from DUAL".

Now the Purchase Order number will automatically populate as soon as the program is launched from Purchase Order form.

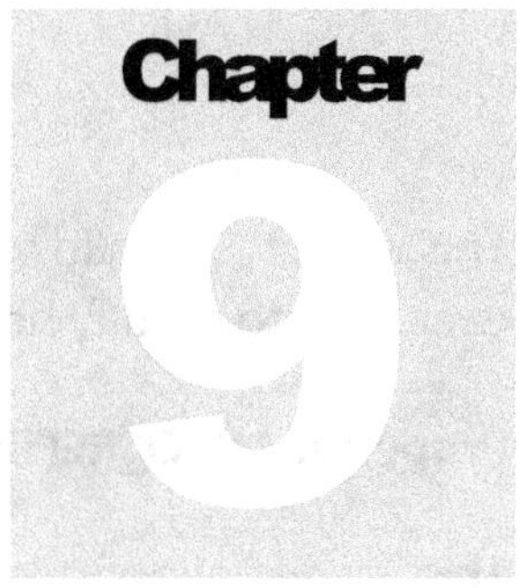

Object Type: LOV (List of Values)

This chapter explains various Personalization properties available for the "LOV" object type. You will be able to change various LOV characteristics using the properties discussed in this chapter.

This object type allows you to change the underlying record group of a LOV in addition to providing the capability to modify height, title, width and horizontal and vertical position.

Property: AUTO_REFRESH

This is a very important property, as setting up this property as "TRUE" allows the related LOV to refresh data every time the LOV is opened; therefore users do not need to close and re-open the form to refresh the LOV data.

We will explain this property in detail using the Line Type LOV of the requisition line.

The Line Type LOV actually lists all the active PO line types configured in the system as below.

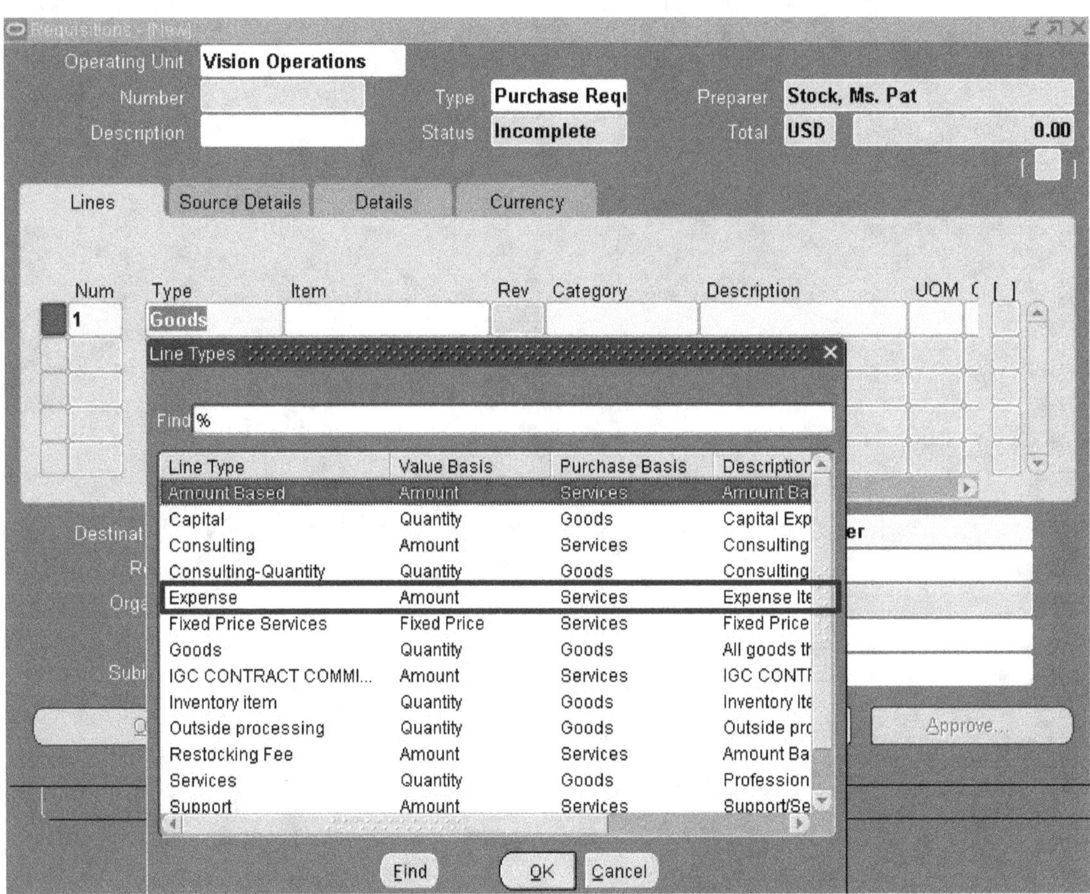

Currently the AUTO_REFRESH property is set up as TRUE. Now if we end-date/deactivate the "Expense" line type and re-open the LOV without closing and re-opening the form, the "Expense" line type will disappear, as the LOV has been auto-refreshed as per below screen shot.

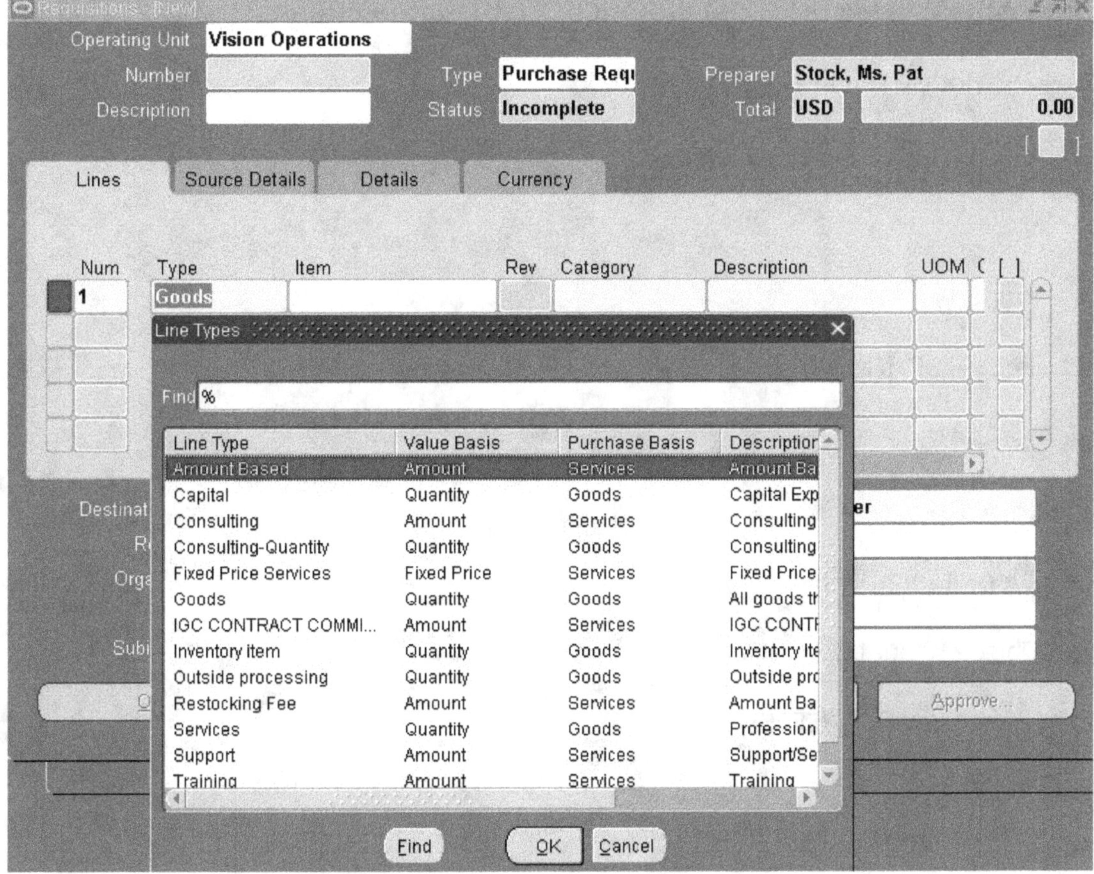

The "Expense" line type doesn't appear in the above LOV.

Had the AUTO_REFRESH property been setup as FALSE, you had to close and re-open the form to see the above mentioned change.

You might come across LOVs that are having the AUTO_REFRESH property value as FALSE. Following personalization steps will help you to change the same to TRUE.

- Open the Forms Personalization window by clicking on the Help -> Diagnostics -> Custom Code -> Personalize menu option.
    - Seq – 10
    - Description – <Enter meaningful Description>
    - Level – Form
    - Enabled – Yes
- Condition:

- Trigger Event - WHEN-NEW-FORM-INSTANCE
- Processing Mode – Not in Enter-Query Mode

- Actions:
    - Seq – 10
    - Type – Property
    - Description – <OPTIONAL>
    - Language – All
    - Enabled – Yes
    - Object Type – LOV
    - Target Object- <Your LOV Name>
    - Property Name: AUTO_REFRESH
    - Value – TRUE

Save the record.

Property: GROUP_NAME

This property allows you to assign a different record group to a LOV replacing the seeded one. This is frequently used to restrict the values appear in the LOVs based on user roles, responsibilities and business requirements.

We will use the line type LOV of the requisition form again and assume that the business wants users to create the requisitions with "Goods" line type only. You might end-date/deactivate the line types except "Goods" but the issue is, the end-dated line types will be unavailable even in the other areas, such as Purchase orders. Therefore you need to change the underlying record group of the line type LOV in the requisition form, so that the restriction applies to the requisition form only.

<u>High level sequence of steps:</u>

1. Identify the .fmb file and open the form using Oracle Form Builder.
2. Locate the LOV and respective record group and copy the SQL attached to the record group.

3. Create a new record group using Form Personalization (explained below) modifying the copied SQL. It's very important not to change the column sequence and numbers, as you might need to modify the conditions in the WHERE clause only.

4. Modify the GROUP_NAME property of the respective LOV using the Forms Personalization (explained below) to map the record group created above to the LOV.

<u>Forms Personalization – explaining Step-3 mentioned above:</u>

In this step we will create a new record group that will show the "Goods" lien type only.

- Open the Forms Personalization window by clicking on the Help -> Diagnostics -> Custom Code -> Personalize menu option.
    - Seq – 10
    - Description – <Enter meaningful Description>
    - Level – Form
    - Enabled – Yes
- Condition:
    - Trigger Event - WHEN-NEW-ITEM-INSTANCE
    - Trigger Object – LINES.LINE_TYPE
    - Processing Mode – Not in Enter-Query Mode
- Actions:
    - Seq – 10
    - Type – Builtin
    - Description – <OPTIONAL>
    - Language – All
    - Enabled – Yes
    - Builtin Type – Create Record Group from Query

- Argument – SELECT line_type, displayed_order_type, purchase_basis_dsp, description FROM po_line_types_v WHERE line_type = 'Goods' (SQL may need to be modified further).
- Group Name: XX_LINE_TYPE

Save the record.

Now a new record group has been created as XX_LINE_TYPE and next step is to attach it to the Line Type LOV as explained below.

<u>Forms Personalization – explaining Step-4 mentioned above:</u>

We will attach the newly created record group to the line type LOV based on the following steps.

- Open the Forms Personalization window by clicking on the Help -> Diagnostics -> Custom Code -> Personalize menu option.
    - Seq – 10
    - Description – <Enter meaningful Description>
    - Level – Form
    - Enabled – Yes
- Condition:
    - Trigger Event - WHEN-NEW-ITEM-INSTANCE
    - Trigger Object – LINES.LINE_TYPE
    - Processing Mode – Not in Enter-Query Mode
- Actions:
    - Seq – 10
    - Type – Property
    - Description – <OPTIONAL>
    - Language – All
    - Enabled – Yes

- Object Type – LOV
- Target Object – LINE_TYPE
- Property Name – Group Name
- Value – XX_LINE_TYPE (initial value was LINE_TYPE)

Save the record.

Now, as per the below screen shot, the line type LOV in the requisition line level will show "Goods" line type only.

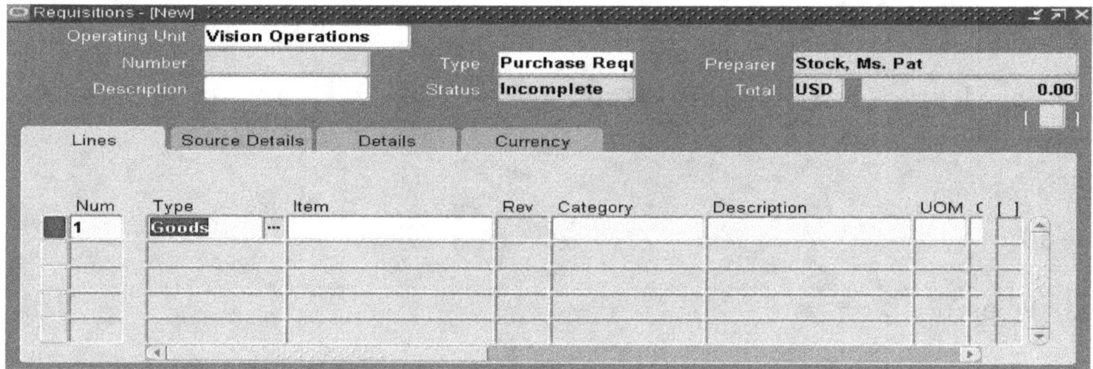

Property: HEIGHT

This property allows you to modify the height of a LOV. We will use the Line Type LOV at the requisition line level to explain this functionality.

Below screen shot shows the usual height of the line type LOV.

Following personalization steps will modify the height of the mentioned LOV.

- Open the Forms Personalization window by clicking on the Help -> Diagnostics -> Custom Code -> Personalize menu option.
    - Seq – 10
    - Description – <Enter meaningful Description>
    - Level – Form
    - Enabled – Yes
- Condition:
    - Trigger Event - WHEN-NEW-FORM-INSTANCE
    - Processing Mode – Not in Enter-Query Mode
- Actions:
    - Seq – 10
    - Type – Property
    - Description – <OPTIONAL>
    - Language – All
    - Enabled – Yes
    - Object Type – LOV
    - Target Object – LINE_TYPE
    - Property Name – HEIGHT
    - Value – 2 (initial value was 4)

Save the record.

Following screen shot shows the modified LOV.

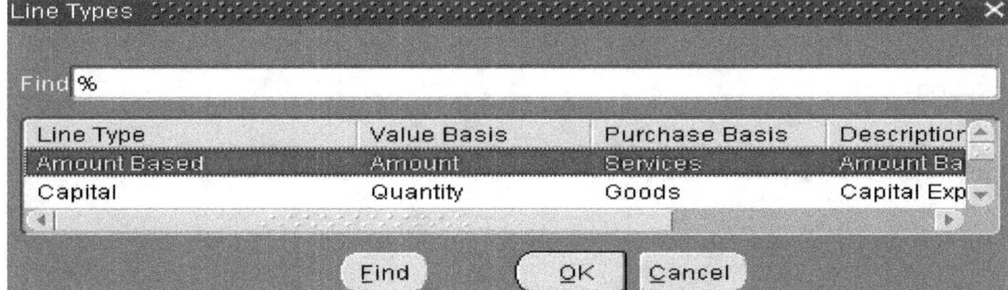

Property: TITLE

You may change the title of a LOV using this property. Actually the title of the Line Type LOV at the requisition line level is “Line Type”. We will change it to “Select an appropriate Line Type” using the below personalization steps.

- Open the Forms Personalization window by clicking on the Help -> Diagnostics -> Custom Code -> Personalize menu option.
    - Seq – 10
    - Description – <Enter meaningful Description>
    - Level – Form
    - Enabled – Yes
- Condition:
    - Trigger Event - WHEN-NEW-FORM-INSTANCE
    - Processing Mode – Not in Enter-Query Mode
- Actions:
    - Seq – 10
    - Type – Property
    - Description – <OPTIONAL>
    - Language – All
    - Enabled – Yes
    - Object Type – LOV

- Target Object – LINE_TYPE
- Property Name – TITLE
- Value – Select an appropriate Line Type

Save the record.

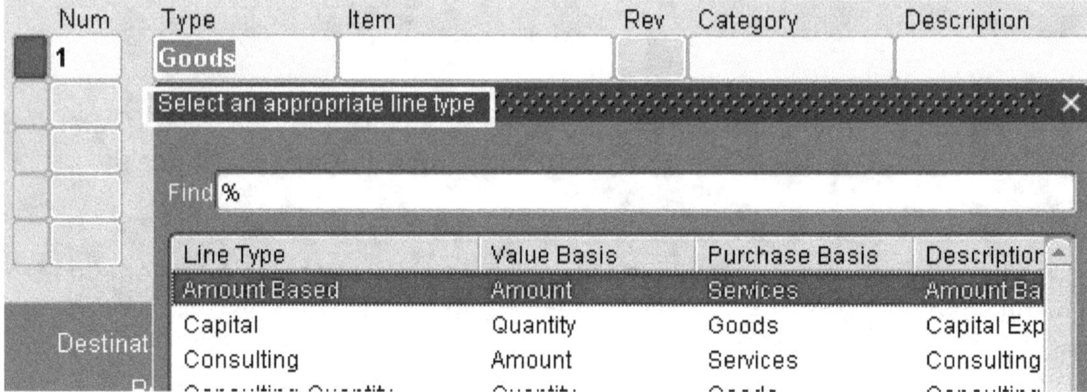

Above screen shot shows the modified title.

Property: WIDTH

As the name suggests, this property will help you to modify the width of a LOV. Numerous screen shots have shown the standard width of the Line Type LOV where we notice that the description of the line type isn't visible fully unless user scrolls right. In order to make the whole content visible to the user, you may perform the following personalization steps to increase the width of the LOV. This will help to increase the user experience and convenience.

- Open the Forms Personalization window by clicking on the Help -> Diagnostics -> Custom Code -> Personalize menu option.
    - Seq – 10
    - Description – <Enter meaningful Description>
    - Level – Form
    - Enabled – Yes
- Condition:
    - Trigger Event - WHEN-NEW-FORM-INSTANCE
    - Processing Mode – Not in Enter-Query Mode

- Actions:
  - Seq – 10
  - Type – Property
  - Description – <OPTIONAL>
  - Language – All
  - Enabled – Yes
  - Object Type – LOV
  - Target Object – LINE_TYPE
  - Property Name – WIDTH
  - Value – 8 (initial value 5)

Save the record.

Modified LOV is shown below.

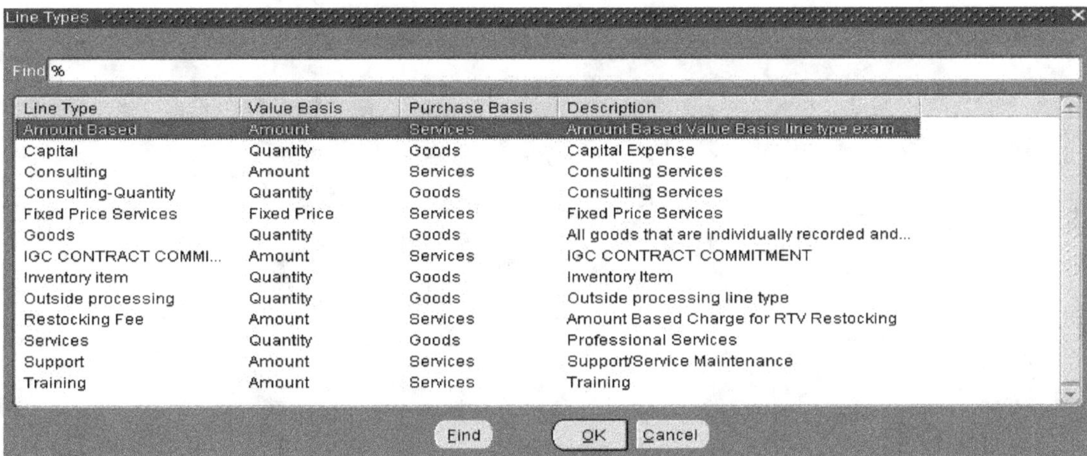

Line Types

Find %

Line Type	Value Basis	Purchase Basis	Description
Amount Based	Amount	Services	Amount Based Value Basis line type exam...
Capital	Quantity	Goods	Capital Expense
Consulting	Amount	Services	Consulting Services
Consulting-Quantity	Quantity	Goods	Consulting Services
Fixed Price Services	Fixed Price	Services	Fixed Price Services
Goods	Quantity	Goods	All goods that are individually recorded and...
IGC CONTRACT COMMI...	Amount	Services	IGC CONTRACT COMMITMENT
Inventory item	Quantity	Goods	Inventory Item
Outside processing	Quantity	Goods	Outside processing line type
Restocking Fee	Amount	Services	Amount Based Charge for RTV Restocking
Services	Quantity	Goods	Professional Services
Support	Amount	Services	Support/Service Maintenance
Training	Amount	Services	Training

Find OK Cancel

Chapter

10

Object Type: Local Variable

This chapter explains various Personalization properties available for the "Local Variable" object type. You will be able to understand the difference between "Local Variable" and "Global Variable" object types in addition to understanding the capability of this object type in this chapter.

Local Variable works similar to a Global Variable (discussed above) but its effectivity is limited to a single form only. It is highly recommended not to use Global Variable when the scope of the business logic is limited to a single form. We will demonstrate the functionality of the Local variable based on the below example.

<u>Business Scenario:</u>

Those who have worked on Oracle Item Master, knows that there are several item categories that gets assigned to an item automatically upon saving the item record. But it is important for the user to open the item category form and update the default categories accordingly, as the default ones are usually generic, such as MISC.MISC or 0000.0000 etc. as explained in the below screen shot.

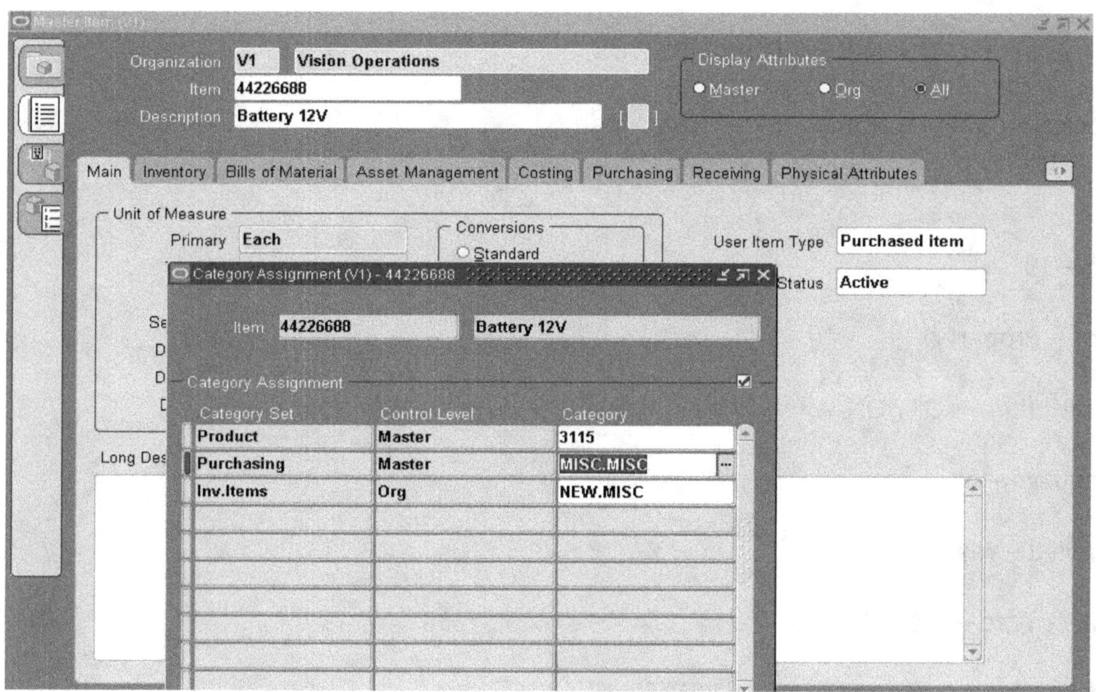

Usually the end-users forget to open the Category Assignment form to update the default category codes, as this is not a mandatory activity. Therefore the business will end up having several items with incorrect category codes leading to incorrect reporting.

Local Variable will enable you to setup a validation mechanism and notify the user accordingly, when they save the item record without opening the category assignment form.

<u>Personalization Step-1:</u>

In this step, we will define a Local Variable and assign the value "1" to it. Idea is to update this value to "2" as soon as the category assignment form is opened (explained in Step-2). An error message will be shown if the value of the Local Variable still remains one while user saves the record (explained in Step-3).

- Open the Forms Personalization window by clicking on the Help -> Diagnostics -> Custom Code -> Personalize menu option.
    - Seq – 10
    - Description – Define and Assign Local Variable
    - Level – Form
    - Enabled – Yes
- Condition:
    - Trigger Event - WHEN-NEW-FORM-INSTANCE
    - Processing Mode – Not in Enter-Query Mode
- Actions:
    - Seq – 10
    - Type – Property
    - Description – <OPTIONAL>
    - Language – All
    - Enabled – Yes
    - Object Type – Local Variable
    - Target Object – XX_CATEGORY_UPDATE
    - Property Name – Value
    - Value – 1

Save the record.

<u>Personalization Step-2:</u>

In this step, we will assign the value "2" to the Local Variable, as soon as the category assignment form/block is opened.

- Add another header row in the Forms Personalization window.

- Seq – 20
- Description – Update the Local Variable assignment upon opening the category assignment form/block
- Level – Form
- Enabled – Yes

- Condition:
  - Trigger Event - WHEN-NEW-BLOCK-INSTANCE
  - Trigger Object – CATEGORY_ASSIGN
  - Processing Mode – Not in Enter-Query Mode

- Actions:
  - Seq – 10
  - Type – Property
  - Description – <OPTIONAL>
  - Language – All
  - Enabled – Yes
  - Object Type – Local Variable
  - Target Object – XX_CATEGORY_UPDATE
  - Property Name – Value
  - Value – 2

Save the record.

Personalization Step-3:

We will establish a validation mechanism in this step and display a warning message to the user if the category assignment form has not been opened prior to saving the record.

- Add another header row in the Forms Personalization window.

- Seq – 30
- Description – Validate and Warn
- Level – Form
- Enabled – Yes

- Condition:
  - Trigger Event - WHEN-VALIDATE-RECORD
  - Trigger Object – MTL_SYSTEM_ITEMS
  - Condition - ${VAR.XX_CATEGORY_UPDATE} = 1
  - Processing Mode – Not in Enter-Query Mode

- Actions:
  - Seq – 10
  - Type – Message
  - Description – <OPTIONAL>
  - Language – All
  - Enabled – Yes
  - Message Type – Show
  - Message Text – You have not reviewed the category codes before saving the item setup.

Save the record.

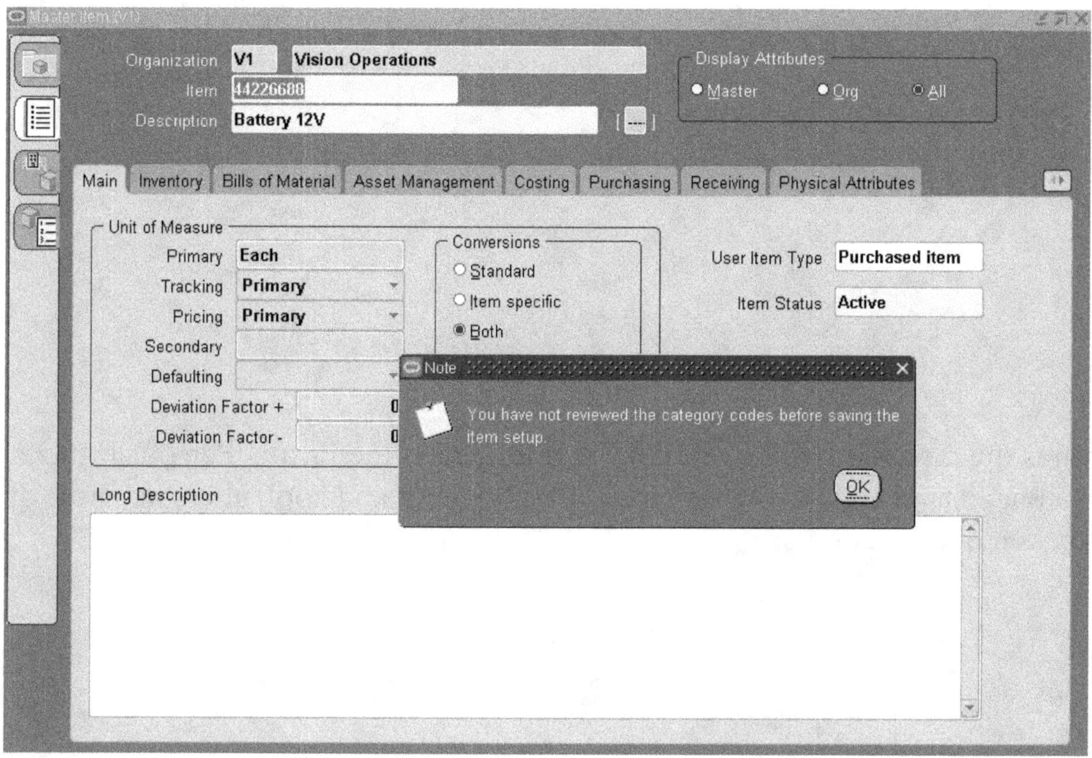

Personalization setup is now complete. Now above warning message appears, if the user saves the item setup data without reviewing the category assignment details.

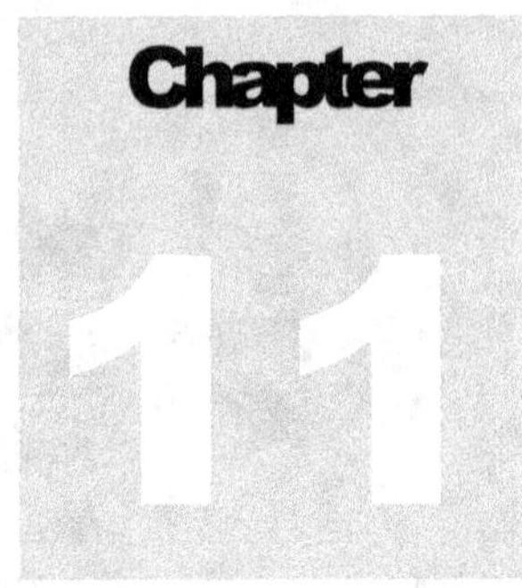

Actions: Builtin

This chapter explains the capability of the "Builtin" action type in the Forms Personalizations. You will be able to understand various builtin types in this chapter and apply them accordingly based on the business requirements.

Builtin Type: Launch SRS Form

The "Launch SRS Form" builtin type allows launching a concurrent program based on the trigger event defined in the "Conditions" tab of the Form Personalizations window. Please refer to "Business Scenario-1" example of the "Object Type: Global Variable" chapter to learn how this builtin type works.

Builtin Type: Launch a Function

The "Launch a Function" builtin type allows launching a form function based on the trigger event defined in the "Conditions" tab of the Form Personalizations window. This is very useful forms personalization mechanism to make additional information available to the user when performing a particular activity in the Oracle Applications. Following business scenario will explain this builtin type.

<u>Business Scenario:</u>

It has been observed in multiple businesses that, purchase requisitions being created by the users to procure item when it is already available in the stock. This actually happens when the requisitions are not created by the automated planning mechanism and the user does not have the visibility to the available stock. This not only leads to overstocking, obsolescence and spoilage etc. but also incurs high inventory carrying cost.

To facilitate informed decision making, it will be very useful if the user has the facility to check stock while creating the requisition.

Following personalization steps will launch the "Material Workbench" (Inventory On-Hand Inquiry) form upon opening the purchase requisition form. User can drag the "Material Workbench" form to the right of the Purchase Requisition form while creating the requisition, so that stock check can be performed as needed.

Responsibility: Purchasing Super User or similar

Navigation: Requisitions -> Requisitions

- Open the Forms Personalization window by clicking on the Help -> Diagnostics -> Custom Code -> Personalize menu option.
    - Seq – 10
    - Description – Launch Material Workbench
    - Level – Form
    - Enabled – Yes
- Condition:

- Trigger Event - WHEN-NEW-FORM-INSTANCE
- Processing Mode – Not in Enter-Query Mode

- Actions:
  - Seq – 10
  - Type – Builtin
  - Description – <OPTIONAL>
  - Language – All
  - Enabled – Yes
  - Builtin Type – Launch a Function
  - Function Code – INV_INVMATWB
  - Function Name – Material Workbench

Save the record.

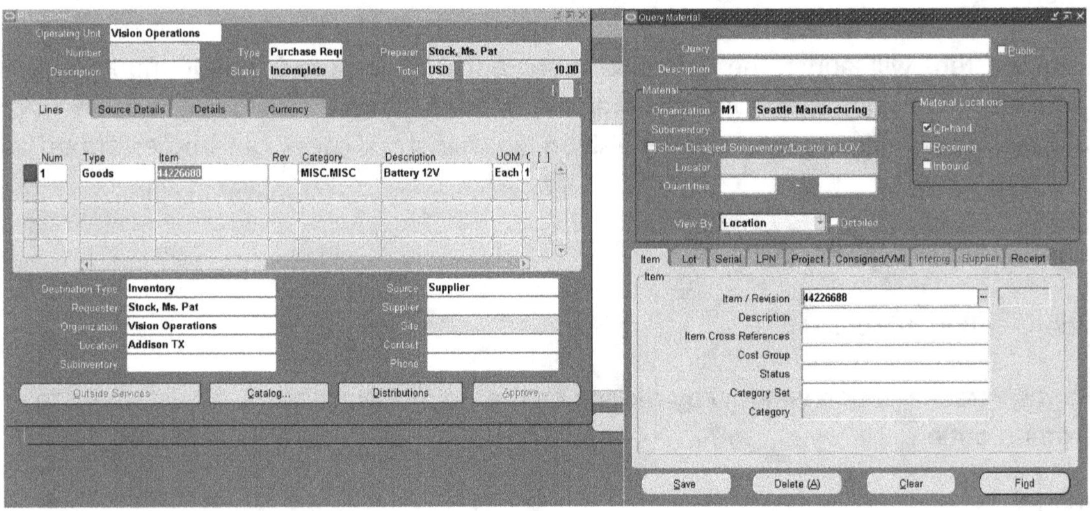

Above screen shot shows the Material Workbench appears alongside the Purchase Requisition form to enable users perform stock checking while creating the requisition.

Builtin Type: Launch a URL

This Builtin Type allows you to launch a URL based on the trigger event defined in the "Conditions" tab of the Form Personalizations window. This is particularly useful if you want to launch a particular webpage (such as- user procedures, data sheets, reference materials etc.) to be launched along with the form to make the useful supporting resources readily available to the user.

Following steps demonstrate the personalization steps.

Responsibility: Purchasing Super User or similar

Navigation: Requisitions -> Requisitions

- Open the Forms Personalization window by clicking on the Help -> Diagnostics -> Custom Code -> Personalize menu option.
    - Seq – 10
    - Description – Launch oraappsguide.com
    - Level – Form
    - Enabled – Yes
- Condition:
    - Trigger Event - WHEN-NEW-FORM-INSTANCE
    - Processing Mode – Not in Enter-Query Mode
- Actions:
    - Seq – 10
    - Type – Builtin
    - Description – <OPTIONAL>
    - Language – All
    - Enabled – Yes
    - Builtin Type – Launch a URL
    - Argument – www.oraappsguide.com

Save the record.

Now as soon as the Requisition form is launched it will open the www.oraappsguide.com website automatically.

Builtin Type: DO_KEY

This builtin type actually performs the keyboard navigation based on the conditions defined in the "Condition" section of the forms personalization window. Appropriate configuration of the DO_KEY arguments enable user to save several mouse clicks leading to much quicker data entry and operations in the Oracle Forms.

Most of the DO_KEY arguments are self-explanatory but I will make an effort to explain them further.

CLEAR_BLOCK: Clears a particular block of a form. This resembles the function of F7 key and very useful when user needs to enter a row into a data block already containing several rows. For example: If an incomplete requisition already contains 100 rows entered earlier by a user and next time the user comes to enter the 101st row, she might need to scroll all the way down and then start entering the new row. In order to save her scrolling down time, if you clear the block using the CLEAR_BLOCK DO_KEY argument at the WHEN-NEW-BLOCK-INSTANCE trigger of the LINES block, it will actually provide her a clean slate to enter the new row.

CLEAR_FORM: This DO_KEY argument clears the whole form and performs the job of the F8 function key.

CLEAR_RECORD: Clears a particular record from a form block. This works similar to the function key F6.

COMMIT_FORM: As the name suggests, this argument will save the record entered/updated into the data block. This works same as CTRL+S key combination and can be used at the WHEN-NEW-ITEM-INSTANCE level of a particular form field to save a record automatically.

COUNT_QUERY: This argument can be used to get the number of rows available in a particular data block. The number appears either on the status bar or in a message box based on the trigger event. In addition to showing the number of record, it also clears the block to allow new row entry. F12 function key resembles the activity of this argument.

CREATE_RECORD: Creates a new line automatically in a form block allowing user entering new record. This works similar to the function key CTRL+DOWN.

DELETE_RECORD: This will delete record from a form block based on the condition defined in the "Condition" section of the forms personalization window. CTRL+UP key combination is the appropriate alternative of this argument.

DUPLICATE_RECORD: This argument will allow user to create a duplicate row in a form block. Please note that, not all the form blocks will allows this entry, as there are several form blocks that maintains uniqueness of the records based on line number, sequence number etc. Shift+F6 key combination also performs this operation.

ENTER_QUERY: Initiates the query mode to allow users enter the query condition. This argument functions same as the F11 key.

EXECUTE_QUERY: This argument executes the query based on the condition entered by the user and works similar to CTRL+F11 key combination.

EXIT_FORM: This does the same job as F4 key and closes the form based on the condition defined in the "Condition" section of the forms personalization window.

HELP: This is a very useful DO_KEY argument, as it launches the context sensitive help based on the condition defined in the "Condition" section of the forms personalization window. Users who are working on a form for the first time will find the content of the context sensitive help very useful to learn the relevant procedure at the earliest. CTRL+H key combination does the same.

LIST_VALUES: This argument types will show the list of values automatically when it is called from a WHEN-NEW-ITEM-INSTANCE trigger of a form field item that has a list of values attached to it. For example, the line type field in the PO requisition line block has a list of values containing all the line types attached to it and following personalization steps will automatically show the list as soon as user navigates into the line type field.

Responsibility: Purchasing Super User or similar

Navigation: Requisitions -> Requisitions

- Open the Forms Personalization window by clicking on the Help -> Diagnostics -> Custom Code -> Personalize menu option.
    - Seq – 10
    - Description – List Values from the Line type LOV
    - Level – Form
    - Enabled – Yes
- Condition:
    - Trigger Event - WHEN-NEW-ITEM-INSTANCE
    - Trigger Object – LINES.LINE_TYPE
    - Processing Mode – Not in Enter-Query Mode

- Actions:
    - Seq – 10
    - Type – Builtin
    - Description – <OPTIONAL>
    - Language – All
    - Enabled – Yes
    - Builtin Type – DO_KEY
    - Argument – LIST_VALUES

Save the record.

The list appears as below.

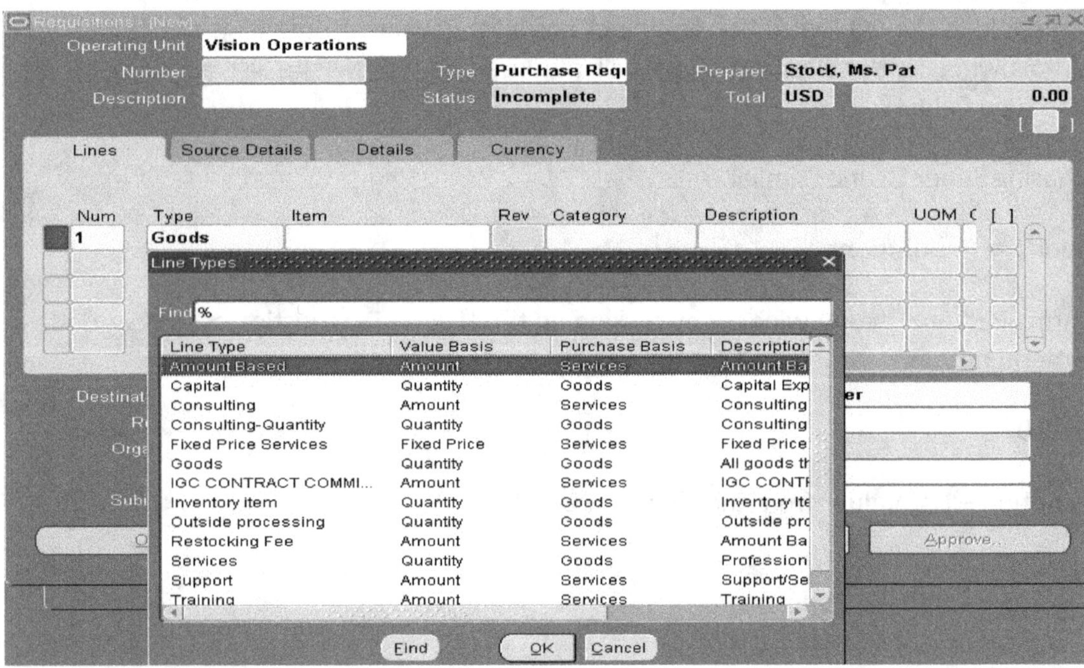

LOCK_RECORD: This DO_KEY argument locks a record which effectively stops any further update to it and shows the below error message.

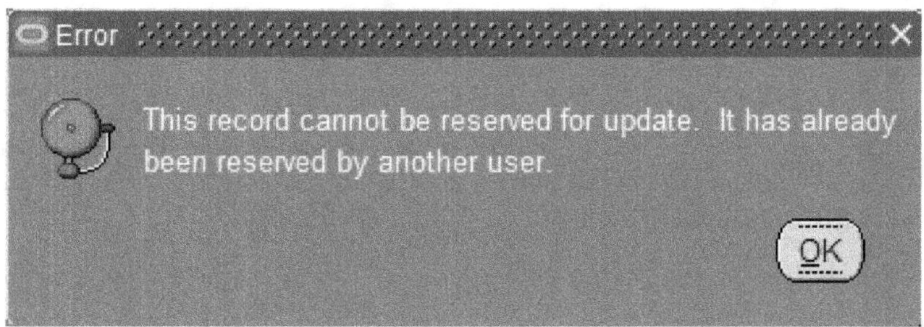

We all are much familiarized with this error message, as this usually appears when a record is being worked on by multiple users at the same time.

NEXT_BLOCK: As the name suggests, this argument moves the cursor to the next block of a form. For example, if it is called from the header block of a form, then the cursor goes to the line block and so on. Shift+PageDown key combination does the same activity.

NEXT_ITEM: This argument follows the TAB key sequence and places the mouse cursor to the next available form field accordingly. If it is called from WHEN-NEW-ITEM-INSTANCE trigger of a form field to place the cursor to the next available form field, effectively you are restricting users to edit/enter values to the field where the WHEN-NEW-ITEM-INSTANCE trigger was fired. This is very useful to guide the users when entering data into a form.

NEXT_RECORD: This argument will place mouse cursor to the next record on a form block. This works similar to a Down Arrow key.

NEXT_SET: It follows the Page Down key activity and scrolls down accordingly. This is required when a form block (usually the Line level ones) contains several rows and you want to show a subset of records to the users straightaway instead of them scrolling all the way down.

PREVIOUS_BLOCK: As the name suggests, this argument moves the cursor to the previous block of a form. For example, if it is called from the line block of a form, then the cursor goes to the header block and so on. Shift+PageUp key combination does the same activity.

PREVIOUS_ITEM: This argument will place the mouse cursor to the previous navigable item. It works similar to a Shift+TAB key combination.

PREVIOUS_RECORD: This works same way as the Up Arrow key and scrolls upwards on a form block to place mouse cursor on the previous record.

PRINT: This DO_KEY argument launches the Print dialog box (as shown below) for users to initiate the printing activity. Ctrl+P key combination is the appropriate alternative for this activity.

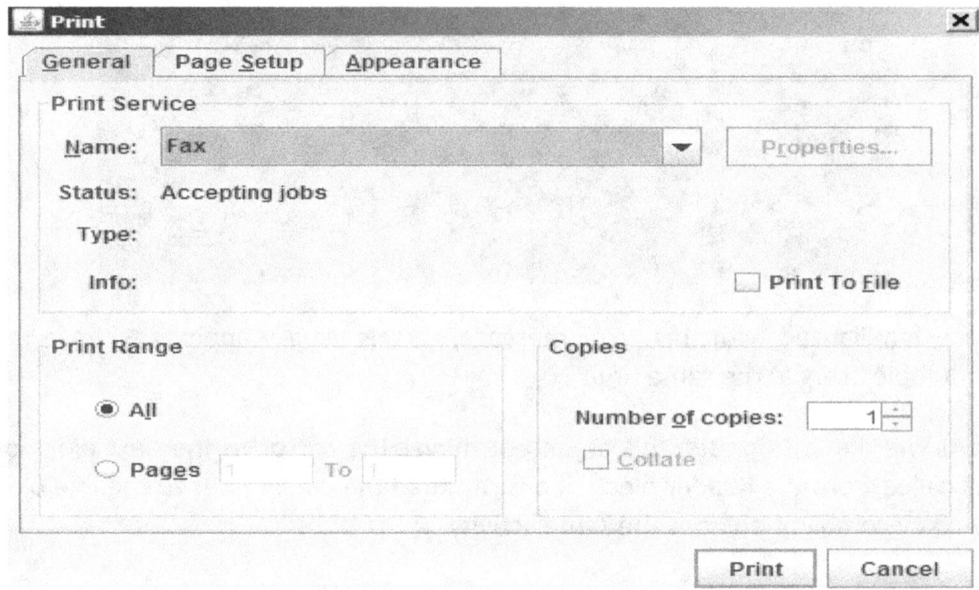

SCROLL_DOWN: This argument is similar to the NEXT_SET argument discussed earlier. This also performs the activity of the Page Down key.

SCROLL_UP: This resembles the activity of the Page Up key and scrolls upwards in a data block.

UP: This DO-KEY argument type places the mouse cursor at the first record in a form block regardless of the current position of the mouse cursor. This will be useful when the mouse cursor is located way down in the form block and user needs to go back to the first record in the block.

Builtin Type: Execute a Procedure

This builtin type extends the capability of Oracle e-Business Suite immensely. You can perform several business critical activities by executing a stored procedure while user is working on a form. Following syntax can be used in the "Argument" field when using the "Execute a Procedure" builtin type.

='declare

begin

PACKAGE.PROCEDURE_NAME('''||${item.mtl_system_items.inventory_item_mir.value}||''');

end'

In the above example, value entered into the "Item" field of the "Master Item" form being used as a parameter to execute a procedure.

You may replace *('''||${item.mtl_system_items.inventory_item_mir.value}||''')* with *(variable_name)* if you are passing a variable as a parameter.

Builtin Type: GO_ITEM

You can control the mouse cursor location or data entry sequence of a user by using this builtin type, as this type allows placing the mouse cursor directly on a form field item based on the trigger and condition defined in the "Condition" tab of the form personalization window. This works similar to NEXT_ITEM and PREVIOUS_ITEM DO_KEY builtin type arguments but those DO_KEY arguments follow the TAB sequence to move the mouse cursor but using the GO_ITEM builtin type you will be able to skip multiple fields and place the mouse cursor on a form field to enable data entry.

We will demonstrate this forms personalization using the Purchase Requisition form accessible through below responsibility and navigation.

Responsibility: Purchasing Superuser or similar

Navigation: Requisitions -> Requisitions

When you open the requisition form, usually the mouse cursor is placed on the "Operating unit" field. Following personalization will enable the mouse cursor to be placed on the "Description" field, if the field does not contain any values while launching the form.

- Open the Forms Personalization window by clicking on the Help -> Diagnostics -> Custom Code -> Personalize menu option.
  - Seq – 10
  - Description – Using GO_ITEM builtin type
  - Level – Form
  - Enabled – Yes
- Condition:
  - Trigger Event - WHEN-NEW-FORM-INSTANCE
  - Condition - ${item.po_req_hdr.description.value} IS NULL
  - Processing Mode – Not in Enter-Query Mode
- Actions:
  - Seq – 10

- Type – Builtin
- Description – <OPTIONAL>
- Language – All
- Enabled – Yes
- Builtin Type – GO_ITEM
- Argument – PO_REQ_HDR.DESCRIPTION

Save the record.

Builtin Type: GO_BLOCK

This builtin type allows placing the mouse cursor directly on a form block based on the trigger and condition defined in the "Condition" tab of the form personalization window. It helps save several key strokes, if there is no data entry requirement on a form block where the cursor is placed and user is using only keyboard navigation to move the cursor from one field to another.

For example, there are many organizations that generate the requisition number automatically and do not encourage entering the requisition description on the header block of the requisition form. Therefore it will be efficient, if the mouse cursor is directly placed on the requisition lines block when launching the form.

Following personalization steps will help you achieve the same.

- Open the Forms Personalization window by clicking on the Help -> Diagnostics -> Custom Code -> Personalize menu option.
    - Seq – 10
    - Description – Using GO_BLOCK builtin type
    - Level – Form
    - Enabled – Yes
- Condition:
    - Trigger Event - WHEN-NEW-FORM-INSTANCE
    - Processing Mode – Not in Enter-Query Mode
- Actions:

- Seq – 10
- Type – Builtin
- Description – <OPTIONAL>
- Language – All
- Enabled – Yes
- Builtin Type – GO_BLOCK
- Argument – LINES

Save the record.

Builtin Type: FORMS_DDL

This builtin type can be used if you want to insert or update record into a table based on the trigger and condition defined in the “Condition” tab of the form personalization window. Following syntax can be used in the “Argument” field when using the inserting records into a table using the “FORMS_DDL” builtin type.

```
='declare

begin

INSERT INTO TABLE_NAME (COLUMN_NAME) VALUES
('''||${item.mtl_system_items.inventory_item_mir.value}||''');

COMMIT;

end'
```

In the above example, value entered into the “Item” field of the “Master Item” form will be inserted into a table.

You may replace *('''||${item.mtl_system_items.inventory_item_mir.value}||''')* with *(variable_name)* if you inserting a value stored in a variable.

Builtin Type: RAISE FORM_TRIGGER_FAILURE

This builtin type is frequently used when activities on a form needs to be suspended based on data entry validation error. Effective use of this builtin type restricts incorrect data entry, as user needs to make necessary corrections to proceed further.

Builtin Type: EXECUTE_TRIGGER

This is a very useful built type that allows you to perform additional activity by executing various seeded and custom triggers. In the following example we will explore how this can be used to reduce unnecessary user clicks to improve data entry timing and user experience.

When we create requisitions using the requisition form available in the below responsibility and navigation, it is required to submit the requisition for approval by clicking on the "Approve" button.

Responsibility: Purchasing Superuser or similar

Navigation: Requisitions -> Requisitions

Below form is presented upon clicking on the "Approve" button and user needs to click on the "OK" button to submit the requisition for approval.

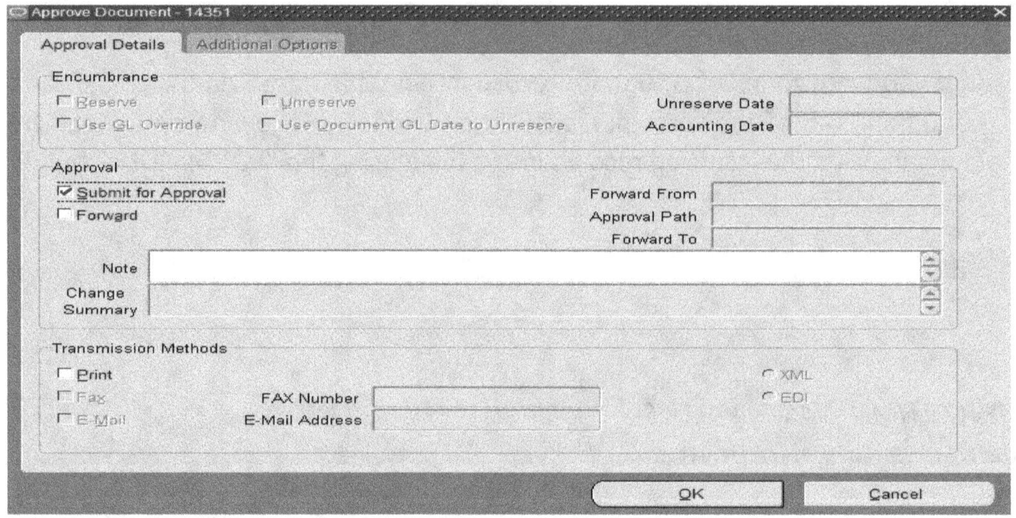

Several organizations advise users just to submit the requisition for approval without making any changes in the approval form. Therefore for them, user coming to this form and clicking on the "OK" button is a wasted effort. This wasted effort becomes significant when users are creating several requisitions a day.

Below personalization steps will submit a requisition for approval without presenting the "Approve Document" form, once user clicks on the "Approve" button available in the requisition form. Effectively the personalization will click on the "OK" button on behalf of the user.

- Open the Forms Personalization window by clicking on the Help -> Diagnostics -> Custom Code -> Personalize menu option.
    - Seq – 10
    - Description – Using EXECUTE_TRIGGER builtin type

- Level – Form
- Enabled – Yes

- Condition:
  - Trigger Event - WHEN-NEW-BLOCK-INSTANCE
  - Trigger Object – PO_APPROVE
  - Processing Mode – Not in Enter-Query Mode

- Actions:
  - Seq – 10
  - Type – Builtin
  - Description – <OPTIONAL>
  - Language – All
  - Enabled – Yes
  - Builtin Type – GO_ITEM
  - Argument – PO_APPROVE.WF_OK_BUTTON

  - Seq – 20
  - Type – Builtin
  - Description – <OPTIONAL>
  - Language – All
  - Enabled – Yes
  - Builtin Type – EXECUTE_TRIGGER
  - Argument – WHEN-BUTTON-PRESSED

Save the record.

Builtin Type: Call Custom Library

This builtin type allows you to call custom code available either in the form libraries (.pll files) or in the custom.pll file. The capability allows performing additional activities including validations based on the trigger and condition defined in the "Condition" tab of the form personalization window.

Builtin Type: Create Record Group from Query

This is a very useful builtin type that allows you to modify a LOV based on the business requirements. Please refer to "Property: Group Name" in the "Object Type: LOV" chapter to find the example that explains the capability of this builtin type.

Builtin Type: Set Profile Value in Cache

We have come across many profile options that is usually setup at the site/responsibility/user level, but at times there are requirements that a profile option value needs to be changed temporarily to provide more efficiency to the user. The builtin type "Set Profile Value in Cache" provides the capability to temporarily setup a value for a particular profile to supersede the site/responsibility/user level value. Below example explains the capability.

Example: If we setup "RCV: Print Receipt Traveler" profile option as "Yes" at the site/responsibility/user level, the Receipts form accessible through Oracle Inventory and Purchasing responsibilities will launch "Receipt Traveler" concurrent program automatically once the receipt is created. This program output can be printed automatically if the profile option "Concurrent:Report Copies" is setup with the value 1 (anything greater than 0 will do).

Most of the businesses prefer to maintain the value of the profile option "Concurrent:Report Copies" as 0 to restrict automatic printing but would like to change the value to 1 temporarily just to allow auto printing of the "Receipt Traveler" report.

Following personalization steps will setup the value of the "Concurrent:Report Copies" profile option as 1 as soon as Receipt form is launched using the below responsibility and navigation.

Responsibility: Purchasing or Inventory Superuser or similar.

Navigation: Receiving -> Receipts (Purchasing responsibility) or Transactions -> Receiving -> Receipts (Inventory responsibility)

- Open the Forms Personalization window by clicking on the Help -> Diagnostics -> Custom Code -> Personalize menu option.
    - Seq – 10
    - Description – Setting up profile value in cache

  - Level – Form
  - Enabled – Yes
- Condition:
  - Trigger Event - WHEN-NEW-FORM-INSTANCE
  - Processing Mode – Not in Enter-Query Mode
- Actions:
  - Seq – 10
  - Type – Builtin
  - Description – <OPTIONAL>
  - Language – All
  - Enabled – Yes
  - Builtin Type – Set Profile Value in Cache
  - Profile Name – CONC_COPIES (This is the internal name)
  - Profile Value - 1

Save the record.

Now the report copies will be automatically printed once the "Receipt Traveler" program is completed.

Note: Once the profile value is setup in the cache, this will remain valid until the user logs out of the application. Therefore appropriate care may be taken to reset the profile option back to initial value (e.g.0) by writing another piece of personalization at the appropriate trigger event.

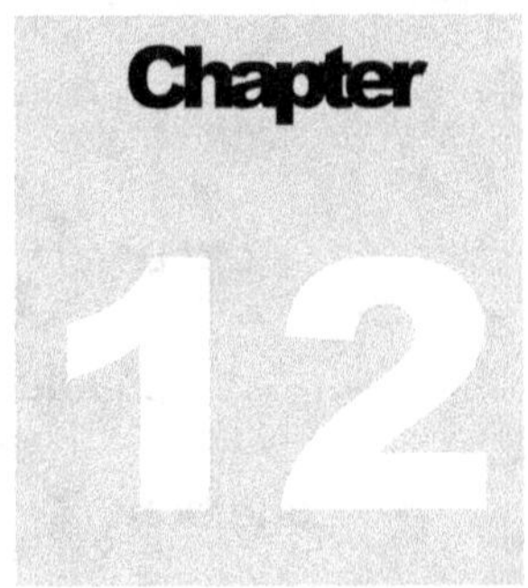

Actions: Message

This chapter explains the "Message" action type. You will be able to prompt various messages applying the Forms Personalization techniques discussed in this chapter.

The “Message” action type is used to display meaningful and relevant messages to users to facilitate effective data entry into the form by showing appropriate notes, warning, hints etc. This action type is also used by the developers to debug technical issues related to form personalizations.

There are 5 message types as – Show, Hint, Error, Debug and Warn. We will explain all these message types in this chapter.

Message Type: Show

You can show a message to the users based on the trigger and condition defined in the “Condition” tab of the form personalization window.

Following personalization steps will show a message to the users as soon as the Requisition form is launched using below responsibility and navigation.

Responsibility: Purchasing Superuser or similar

Navigation: Requisitions -> Requisitions

- Open the Forms Personalization window by clicking on the Help -> Diagnostics -> Custom Code -> Personalize menu option.
    - Seq – 10
    - Description – Working with SHOW message type
    - Level – Form
    - Enabled – Yes
- Condition:
    - Trigger Event - WHEN-NEW-FORM-INSTANCE
    - Processing Mode – Not in Enter-Query Mode
- Actions:
    - Seq – 10
    - Type – Message
    - Description – <OPTIONAL>
    - Language – All
    - Enabled – Yes
    - Message Type – Show

- Message Text – Requisition number will be auto generated once the data is saved.

Save the record.

The message appears as below.

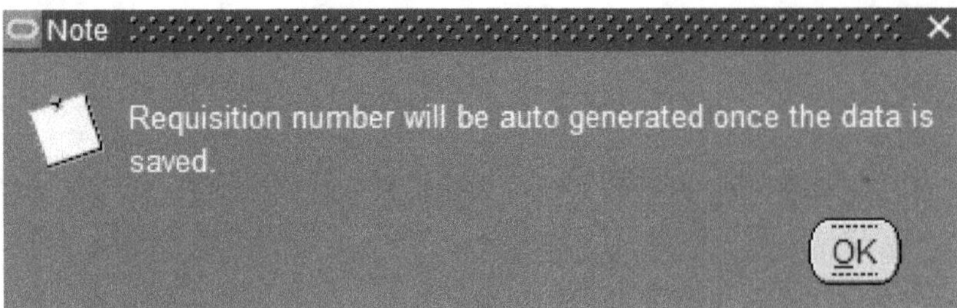

The icon will be different for each message type. Only the "Show" and "Debug" message types share above icon.

Message Type: Hint

This message type can be used as a user guide, as the message appears as a hint at the bottom left corner (status bar) of the form. Many organizations generate this message at the WHEN-NEW-ITEM-INSTANCE trigger, so that users can see relevant message at the status bar as they navigate through the form field.

Following personalization steps will show a hint message when the cursor is placed at the line type field of the requisition form accessible through below responsibility and navigation.

Responsibility: Purchasing Superuser or similar

Navigation: Requisitions -> Requisitions

- Open the Forms Personalization window by clicking on the Help -> Diagnostics -> Custom Code -> Personalize menu option.
    - Seq – 10
    - Description – Working with HINT message type
    - Level – Form
    - Enabled – Yes
- Condition:
    - Trigger Event - WHEN-NEW-ITEM-INSTANCE
    - Trigger Object – LINES.LINE_TYPE
    - Processing Mode – Not in Enter-Query Mode

- Actions:
  - Seq – 10
  - Type – Message
  - Description – <OPTIONAL>
  - Language – All
  - Enabled – Yes
  - Message Type – Hint
  - Message Text – Please select appropriate line type.

Save the record.

Below screen shot shows the hint. You can setup this message type for every user enterable field to show helpful messages.

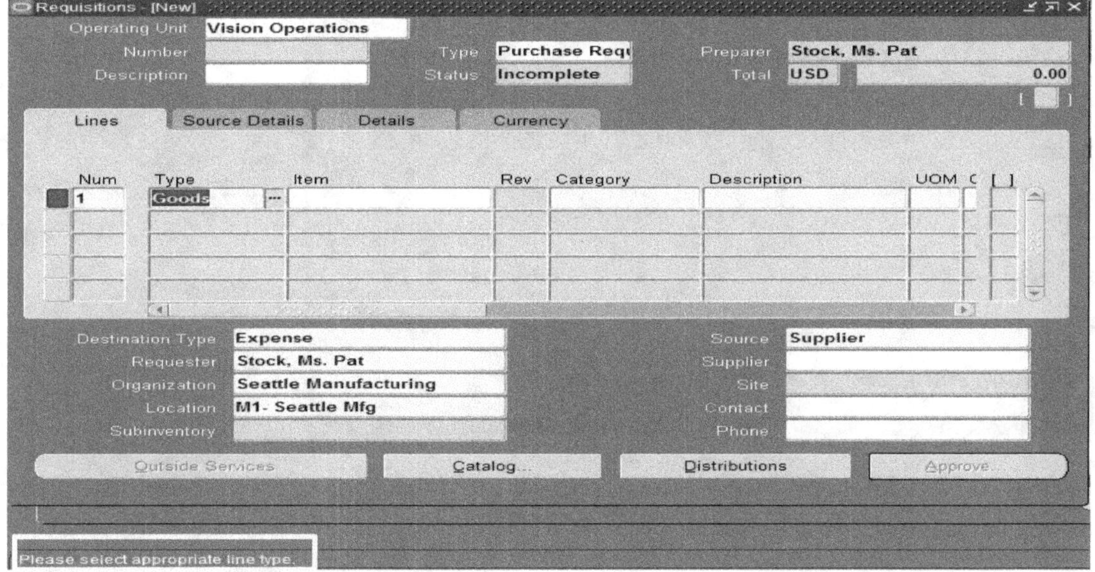

Message Type: Error

As the name suggests, the "Error" message type is used to generate various error messages to notify users while entering data into the forms. An error message appears as below.

The icon is different to the message types discussed earlier.

Message Type: Debug

This message type is available only when the debug mode is setup as "Show Debug Messages" in the forms personalization window.

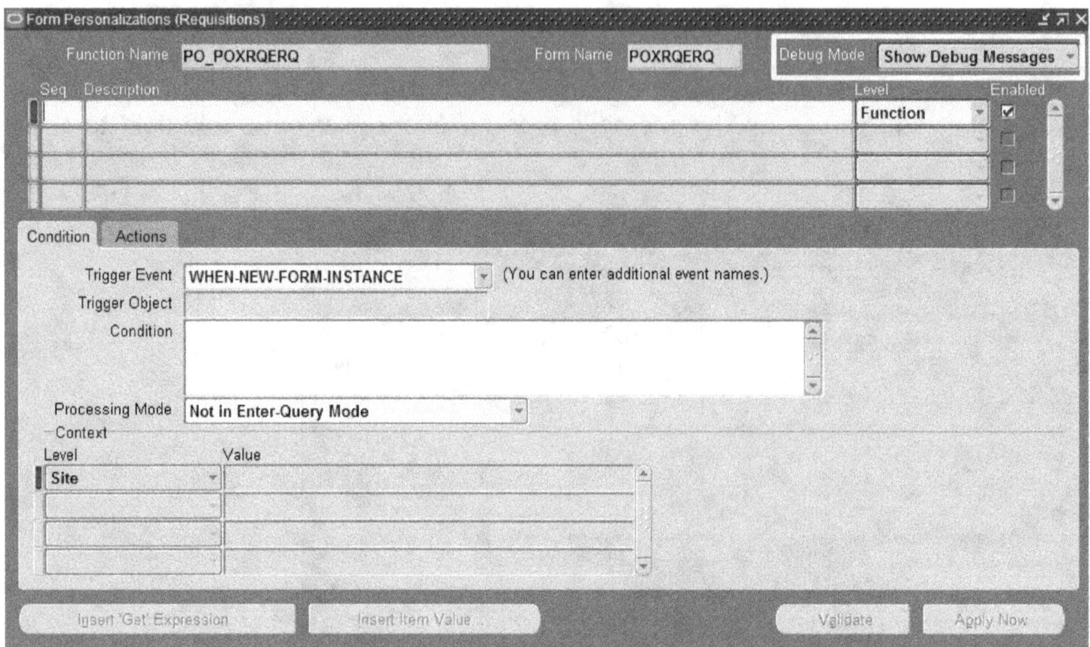

The "Debug" message type is very helpful for the developers to identify the root causes of various forms personalization issues and the message appears as below.

As mentioned earlier, The "Show" and "Debug" messages share the same icon.

Message Type: Warn

The "Warn" message type is used to show a warning message to the user. Following personalization steps will show a warning message when the item field in the requisition form will be left blank. The validation will be performed and warning message issued as soon as the mouse cursor enters the category field.

Responsibility: Purchasing Superuser or similar

Navigation: Requisitions -> Requisitions

- Open the Forms Personalization window by clicking on the Help -> Diagnostics -> Custom Code -> Personalize menu option.
    - Seq – 10
    - Description – Working with WARN message type
    - Level – Form
    - Enabled – Yes
- Condition:
    - Trigger Event - WHEN-NEW-ITEM-INSTANCE
    - Trigger Object – LINES.ITEM_CATEGORY
    - Condition - ${item.lines.item_number.value} IS NULL
    - Processing Mode – Not in Enter-Query Mode
- Actions:
    - Seq – 10
    - Type – Message
    - Description – <OPTIONAL>
    - Language – All
    - Enabled – Yes
    - Message Type – Warn
    - Message Text – The item number field is blank.

Save the record.

The warning message appears as below.

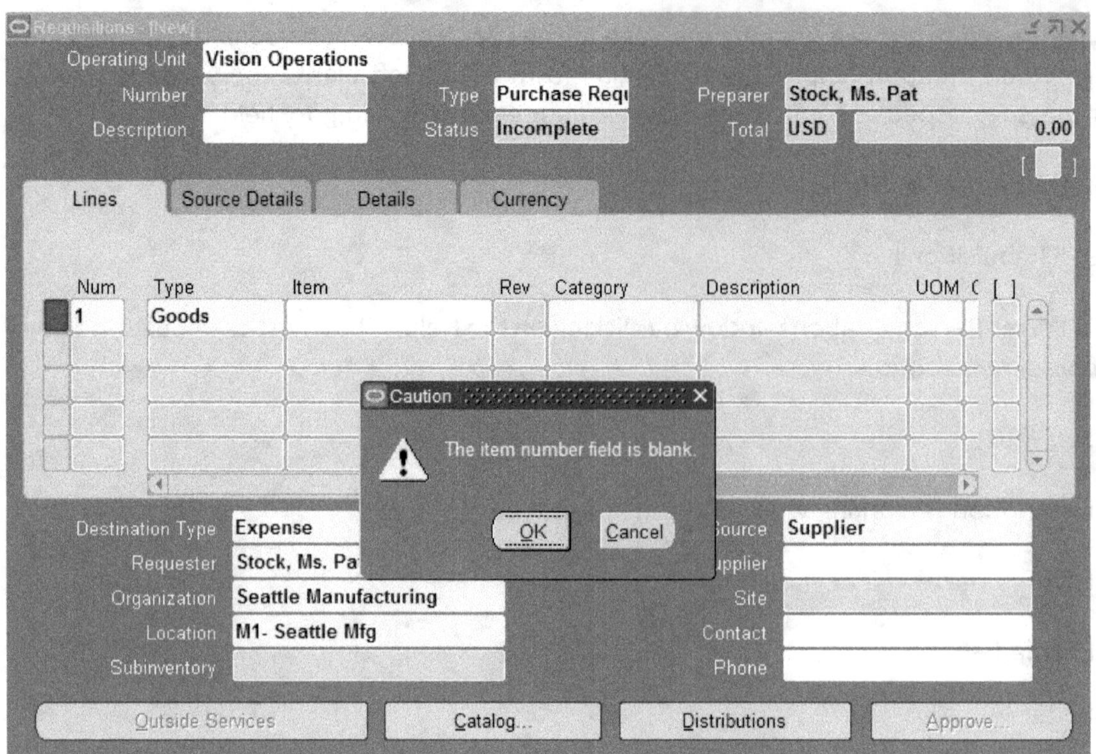
Operating Unit
Vision Operations
Number
Type
Purchase Requ
Preparer
Stock, Ms. Pat
Description
Status
Incomplete
Total
USD
0.00
Lines
Source Details
Details
Currency
Num
Type
Item
Rev
Category
Description
UOM
1
Goods
Caution
The item number field is blank.
OK
Cancel
Destination Type
Expense
Requester
Stock, Ms. Pa
Organization
Seattle Manufacturing
Location
M1- Seattle Mfg
Subinventory
Supplier
Site
Contact
Phone
Outside Services
Catalog...
Distributions
Approve...

Conclusion

I sincerely believe that, you have enjoyed reading this book and it has helped you to enhance your knowledge significantly. I encourage you to apply the new piece of knowledge to all your projects to help all the Oracle Applications users across the globe to get an improved experience in their day-to-day operations. Please reach out to me at info@oraappsguide.com if you have any queries in relation to this book. Thank you for reading!!

www.ingramcontent.com/pod-product-compliance
Lightning Source LLC
Chambersburg PA
CBHW081150180526
45170CB00006B/2016
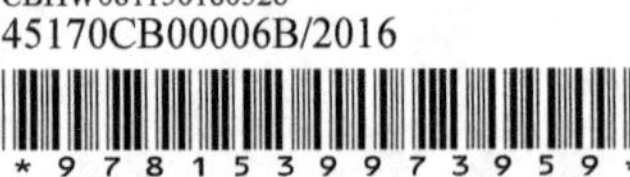